Studies in Computational Intelligence

Volume 493

T0073998

Series Editor

J. Kacprzyk, Warsaw, Poland

For further volumes:
http://www.springer.com/series/7092

Studies in Computational Intelligence

Volume 505

Roger Lee

Editor

Computer and Information Science

 Springer

Editor
Roger Lee
Software Engineering and Information Institute
Central Michigan University
Michigan
USA

ISSN 1860-949X   ISSN 1860-9503   (electronic)
ISBN 978-3-319-00803-5   ISBN 978-3-319-00804-2   (eBook)
DOI 10.1007/978-3-319-00804-2
Springer Cham Heidelberg New York Dordrecht London

Library of Congress Control Number: 2013939056

Printed on acid-free paper

Springer is part of Springer Science+Business Media (www.springer.com)

Preface

The 12th ACIS/IEEE International Conference on Computer Science and Information Science, held in Toki Messe, Nigata, Japan on June 16–20 is aimed at bringing together researchers and scientists, businessmen and entrepreneurs, teachers and students to discuss the numerous fields of computer science, and to share ideas and information in a meaningful way. This publication captures 20 of the conference's most promising papers, and we impatiently await the important contributions that we know these authors will bring to the field.

In chapter 1, Mohamed Redha Sidoumou, Kamal Bechkoum and Karima Benatchba present a study that looks at modeling drivers' behavior with a view to contribute to the problem of road rage. The approach adopted is based on agent technology, particularly multi-agent systems. The simulated model is then used to suggest possible ways of alleviating this societal problem.

In chapter 2, Kai Shi et al. propose a general methodology to extend an existing legacy reactive system with anticipatory ability without re-implementing the whole of the legacy system. They also present three case studies to show how to apply their methodology.

In chapter 3, Dapeng Liu, Shaochun Xu, and Huafu Liu review some research fields of software testing, providing discussion on the issues that have emerged but not clearly represented in previous literature, and try to establish a philosophical viewpoint of understanding software testing. They point out that software testing is intrinsically dynamic and by nature a representative problem of multi-objective optimization.

In chapter 4, Teruhisa Hochin and Hiroki Nomiya propose semantic specialization. This specialization makes it possible for a shape graph, which corresponds to a relation schema in the relational data model, to have elements and edges which are different from those of the original shape graphs, but are semantically related to them.

In chapter 5, Yücel Uğurlu proposes a novel approach to context recognition for use in extracting user interests in e-learning systems. Experimental results show that the proposed approach is robust and able to identify the proper context 96 percent of the time.

In chapter 6, Satoshi Kimura and Hiroyuki Inaba propose a novel visualization system of IDS considering order relation of IP addresses that emphasize the anomalous warning events based on past tendency.

In chapter 7, Insu Song and John Vong present a mobile phone-based banking system, called ACMB (Affective Cashless Mobile Banking), for microfinance. ACMB is designed to provide banking services to the world's unbanked population of 2.2 billion.

In chapter 8, Shuya Ishida et al. propose Shadow detection methods based on new shadow models for object detection. The proposed model includes Normalized Vector Distance instead of Y component. It is a robust feature to illumination changes and can remove a shadow effect in part. The proposed method can obtain shadow regions more accurately by including Normalized Vector Distance in the shadow model.

In chapter 9, Takuya Nakagawa, Yuji Iwahori and M.K. Bhuyan propose a new method to improve the classification accuracy by multiple classes classification using multiple SVM. The proposed approach classifies the true and pseudo defects by adding features to decrease the incorrect classification.

In chapter 10, Paulo Mauricio Goncalves Junior and Roberto Souto Maior de Barros propose an improvement to RCD to perform the statistical tests in parallel by the use of a thread pool and presents how parallelism impacts performance. RCD is a framework for dealing with recurring concept drifts.

In chapter 11, Óscar Mortágua Pereira, Rui L. Aguiar, and Maribel Yasmina Santos propse Concurrent Tuple Set Architecture (CTSA) to manage local memory structures. Their architecture is meant to overcome current fragilities present in Call Level Interfaces.

In chapter 12, Masashi Kawaguchi, Naohiro Ishii and Masayoshi Umeno develop a neuro chip and an artificial retina chip to comprise the neural network model and simulate the biomedical vision system.

In chapter 13, Nomin Batnyam, Ariundelger Gantulga and Sejong Oh propose an efficient method to facilitate SNP data classification. Single Nucleotide Polymorphism (SNP), a unit of genetic variations, has caught much attention recently, as it is associated with complex diseases.

In chapter 14, Tomohiko Takagi and Zengo Furukawa show a novel construction technique of large operational profiles in order to effectively apply statistical software testing to recent projects in which large software is developed in a short timeframe.

In chapter 15, Ahmad Karawash, Hamid Mcheick, and Mohamed Dbouk discuss a model for verifying service composition by building a distributed semi-compiler of service process. They introduce a technique that solves the service composition problems such as infinite loops, deadlock and replicate use of the service.

In chapter 16, James Church et al. look at one aspect of automated species identification: unfolding curved specimens, which commonly occur when specimens are prepared for storage in natural history collections. They examine various techniques for estimating the medial axis of an object, then propose a new method for medial axis estimation based on localized spatial depth.

In chapter 17, Yi Chen et al. describe an upgrade version of MGMR, a pipelined multi-GPU MapReduce system (PMGMR), which addresses the challenge of big data. PMGMR employs the power of multiple GPUs, improves GPU utilization using new GPU features such as streams and Hyper-Q, and handles large data sets which exceed GPU and even CPU memory.

In chapter 18, Shigeaki Tanimoto et al. propose and compile countermeasures to 21 risk factors identified through analysis of risk in a hybrid cloud configuration. With recent progress in Internet services and high-speed network environments, cloud computing has rapidly developed.

In chapter 19, Khan Md Mahfuzus Salam, Takadama Keiki, and Nishio Tetsuro focus on the Pareto principal, which is widely known in the field of Economics. Their motivation is to investigate the reason of why the Pareto principal exists in the human society. They proposed a model for human-agent and conduct simulation.

In chapter 20, Yucong Duan et al. adopt an evolutionary strategy towards approaching a holistic solution for the improvement of the learning efficiency of secondary language (L2) learning.

It is our sincere hope that this volume provides stimulation and inspiration, and that it will be used as a foundation for works to come.

June 2013
Tokuro Matsuo
Program Chair

Contents

List of Contributors

Rui L. Aguiar
DETI–University of Aveiro, Portugal
ruilaa@ua.pt

Yang Bai
Hainan University, China
yang19828@yahoo.com

Henry Bart Jr.
Tulane University, USA
hbartjr@tulane.edu

Nomin Batnyam
Dankook University, South Korea
gngrfish@yahoo.com

Kamal Bechkoum
University of Northampton, UK
kamal.bechkoum@northampton.
ac.uk

Karima Benatchba
Ecole Nationale Supérieure d'Informatique,
Algeria
k_benatchba@esi.dz

M.K. Bhuyan
IIT Guwahati, India
mkb@iitg.ernet.in

Yixin Chen
University of Mississippi, USA
ychen@cs.olemiss.edu

Yi Chen
Arkansas State University, USA
yi.chen@smail.astate.edu

Jingde Cheng
Saitama University, Japan
cheng@aise.ics.saitama-u.ac.jp

James Church
University of Mississippi, USA
jcchurch@go.olemiss.edu

Christophe Cruz
Le2i, CNRS, Dijon, France
christophe.cruz@u-bourgogne.fr

Xin Dang
University of Mississippi, USA
xdang@olemiss.edu

Spencer Davis
Arkansas State University, USA
spencer.davisg@smail.astate.edu

Mohamed Dbouk
Lebanese University,
Rafic-Hariri Campus, Lebanon
mdbouk@ul.edu.lb

Roberto Souto Maior de Barros
Cidade Universitária, Brasil
roberto@cin.ufpe.br

Yucong Duan
Hainan University, China
duanyucong@hotmail.com

Wencai Du
Hainan University, China
wencai@hainu.edu.cn

Abdelrahman Osman Elfaki
MSU, Malaysia
abdelrahmanelfaki@gmail.com

Shinji Fukui
Aichi University of Education, Japan
sfukui@auecc.aichiedu.ac.jp

Zengo Furukawa
Kagawa University, Japan
zengog@eng.kagawa-u.ac.jp

Ariundelger Gantulga
Dankook University, South Korea
ariuka_family@yahoo.com

Paulo Mauricio Gonçalves Júnior
Cidade Universitária, Brasil
paulogoncalves@recife.
   ifpe.edu.br

Yuichi Goto
Saitama University, Japan
gotoh@aise.ics.saitama-u.ac.jp

Teruhisa Hochin
Kyoto Institute of Technology, Japan
hochin@kit.ac.jp

Hiroyuki Inaba
Kyoto Institute of Technology, Japan
inaba@kit.ac.jp

Shuya Ishida
Chubu University, Japan
ishida@cvl.cs.chubu.ac.jp

Naohiro Ishii
Aichi Institute of Technology, Japan
ishii@aitech.ac.jp

Yuji Iwahori
Chubu University, Japan
iwahori@cs.chubu.ac.jp

Motoi Iwashita
Chiba Institute of Technology, Japan
iwashita.motoi@it-chiba.ac.jp

Hai Jiang
Arkansas State University, USA
hjiang@astate.edu

Atsushi Kanai
Hosei University, Japan
yoikana@hosei.ac.jp

Ahmad Karawash
University of Quebec at Chicoutimi,
   Canada
ahmad.karawash1@uqac.ca

Masashi Kawaguchi
Suzuka National College of Technology,
   Japan
masashi@elec.suzuka-ct.ac.jp

Takadama Keiki
The University of Electro-Communications,
   Japan
keiki@inf.uec.ac.jp

Satoshi Kimura
Kyoto Institute of Technology, Japan
kimura08@sec.is.kit.ac.jp

Kuan-Ching Li
Providence University, Taiwan
kuan-cli@pu.edu.tw

Dapeng Liu
GradientX, USA
dapeng@gradientx.com

Huafu Liu
Changsha University, China
hfliu9063@163.com

Shinsuke Matsui
Chiba Institute of Technology,
   Japan
matsui.shinsuke@it-chiba.ac.jp

Hamid Mcheick
University of Quebec at Chicoutimi,
   Canada
hamid_mcheick@uqac.ca

Chihiro Murai
Chiba Institute of Technology, Japan
S0942124KM @it-chiba.ac.jp

Takuya Nakagawa
Chubu University, Japan
tnakagwa@cvl.cs.chubu.ac.jp

Hiroki Nomiya
Kyoto Institute of Technology, Japan
nomiya@kit.ac.jp

Sejong Oh
Dankook University, South Korea
dkumango@gmail.com

Óscar Mortágua Pereira
DETI–University of Aveiro, Portugal
omp@ua.pt

Zhi Qiao
Arkansas State University, USA
zhi.qiao@smail.astate.edu

Khan Md Mahfuzus Salam
The University of Electro-Communications,
   Japan
kmahfuz@gmail.com

Maribel Yasmina Santos
DSI-Univerity of Minho, Portugal
maribel@dsi.uminho.pt

Hiroyuki Sato
The University of Tokyo, Japan
schuko@satolab.itc.
   u-tokyo.ac.jp

Ray Schmidt
Tulane University, USA
rschmidt@tulane.edu

Yosiaki Seki
NTT Secure Platform Laboratories
seki.yoshiaki@lab.ntt.co.jp

Kai Shi
Saitama University, Japan
Northeastern University, China
shikai@aise.ics.saitama-u.ac.jp
shik@swc.neu.edu.cn

Mohamed Redha Sidoumou
University Saad Dahlab of
Blida, Algeria
sidoumoumr@hotmail.com

Insu Song
James Cook University Australia,
   Singapore Campus, Singapore
insu.song@jcu.edu.au

Tomohiko Takagi
Kagawa University, Japan
ftakagi @eng.kagawa-u.ac.jp

Shigeaki Tanimoto
Chiba Institute of Technology, Japan
shigeaki.tanimoto@it-chiba.
   ac.jp

Nishio Tetsuro
The University of Electro-Communications,
   Japan
nishino@uec.ac.jp

Yücel Uğurlu
The University of Tokyo, Japan
yucel.ugurlu@ni.com

Masayoshi Umeno
Chubu University, Japan
umeno@solan.chubu.ac.jp

John Vong
James Cook University Australia,
    Singapore Campus, Singapore
john.vong@jcu.edu.au

Robert J. Woodham
University of British Columbia, Canada
woodham@cs.ubc.ca

Shaochun Xu
Changsha University, China
simon.xu@algomau.ca

Zhiliang Zhu
Northeastern University, China
zhuzl@swc.neu.edu.cn

Drivers' Behaviour Modelling for Virtual Worlds

A Multi-agent Approach

Mohamed Redha Sidoumou, Kamal Bechkoum, and Karima Benatchba

Abstract. In this paper we present a study that looks at modelling drivers' behaviour with a view to contribute to the problem of road rage. The approach we adopt is based on agent technology, particularly multi-agent systems. Each driver is represented by a software agent. A virtual environment is used to simulate drivers' behaviour, thus enabling us to observe the conditions leading to road rage. The simulated model is then used to suggest possible ways of alleviating this societal problem. Our agents are equipped with an emotional module which will make their behaviours more human-like. For this, we propose a computational emotion model based on the OCC model and probabilistic cognitive maps. The key influencing factors that are included in the model are personality, emotions and some social/personal attributes.

Keywords: Agent, emotions, personality, behaviour, simulation.

1 Introduction

Road rage is one of the major problems that we face in today's roads. In this paper we propose to study this phenomenon through simulation, with a view to come up

Mohamed Redha Sidoumou
Computer Science Department, University Saad Dahlab of Blida,
Route Soumaa 09000, Blida, Algeria
e-mail: sidoumoumr@hotmail.com

Kamal Bechkoum
School of Science & Technology, University of Northampton,
Avenue Campus NN2 6JD, Northampton, UK
e-mail: kamal.bechkoum@northampton.ac.uk

Karima Benatchba
Ecole Nationale Supérieure d'Informatique, OUED SMAR, 16309, Algiers, Algeria
e-mail: k_benatchba@esi.dz

R. Lee (Ed.): *Computer and Information Science*, SCI 493, pp. 1–15.
DOI: 10.1007/978-3-319-00804-2_1   © Springer International Publishing Switzerland 2013

with ways of alleviating its negative impact. For the simulation to be effective we need to model drivers' behaviour in an environment similar to the real world where car drivers and other users of the road compete for the same space.

Drivers' behaviour modelling is a complex and multi-faceted-task. Before attempting to model such a behaviour, we need to understand what constitutes human's cognition. We need to know which parameters are influencing our decisions while driving. We also need to know how these parameters are influenced and, indeed, are there any inter-influences between these parameters.

There are many variables that contribute to humans' cognition, but we are limiting this list to the most relevant ones that are believed to have higher influence on drivers' behaviour. For example, cooking skills can be relevant in daily life cognition and some decisions are influenced by it, but we do not see any direct relevance to the driving process. We also need to support the choice of our parameters by borrowing concepts emanating from psychological and social studies. Such parameters include personality, emotional state, driving skills, age, gender and knowledge of the map.

Once the parameters selected, we present the relationship between them and the factors that affect them. For example: how is the personality affecting emotions? How can emotions change over time and what are the effects of all of these variables on the drivers' behaviour?

This modelling will help us create a drivers' behaviour architecture giving us the possibility to explore its implementation using techniques from AI. For this, intelligent agents appear to be a good candidate given their attributes, such as learning and the ability to cooperate.

In the literature we have reviewed, modelling driver behaviour focuses on low level operations like eye movement, lane changing and cognitive reasoning. They do not address the influence of personality and emotions. Our goal is to reduce road rage and we think that personality and emotions are among the most influential factors in road rage.

In addition to these introductory notes the remainder of this paper is organised around the following sections:

Section two introduces the definition of *Road Rage* and the motivations behind this work. The third section presents the variables that influence drivers' behaviour like personality and emotions with literature discussions. In the fourth section we present our emotion model based on the OCC (named after its authors Orthony, Clore and Collins) and a variation of cognitive maps. We give also an example using our model. The fifth section is about modelling the full behaviour. In this section we present our plan into modelling drivers' behaviour using all the parameters we discussed and our emotion model. Concluding remarks are presented in section six.

2   Road Rage

Some disagreements exist among researchers on how to classify a specific reckless driving behaviour as road rage [1]. According to [1] two definitions are used often in research. The first one is: "*an assault with a motor vehicle or other weapon on other vehicles, precipitated by a specific incident*" [2]. The second definition is: "*a deliberate attempt to harm other persons or property arising from an incident involving use of a motor vehicle"* [3]. According to [1] the NHTSA (National Highway Traffic Safety Administration) clearly distinguishes aggressive driving from road rage. For example, from a legal point of view, aggressive driving like speeding is a traffic offense, whereas road rage is a criminal offense [1]. Takaku also quotes a study conducted by AAA's (American Automobile Association) Foundation for Traffic Safety that found that road-rage incidents increased by more than 50% between 1990 and 1996. The study also quotes the cost of the society in the region of $250 billion per year, in addition to the human casualties [4],[5]. A study carried out by AAA Foundation looked at more than 10,000 road rage incidents committed over seven years, and found they resulted in at least 218 fatalities and another 12,610 injury cases. According to this study, 2.18% of road rages end with death and a great deal of injury cases. From these studies it is pretty clear that road rage is a serious problem that cannot be ignored. The aim of this work is to try and contribute towards minimizing the impact of this problem. To do this we propose to simulate cars' traffic using a number a parameters that influence road rage. The next section outlines some of these influencing parameters.

3   Driver Behaviour Modelling: Influencing Factors

To model drivers' behaviour we start by selecting the different parameters affecting it. Those parameters can be psychological, Social or personal. For the purpose of this work only the parameters that are likely to influence drivers' behaviour are explored. We assume that our drivers have a minimum level of driving capability and therefore driver's technical ability is not taken into account.

3.1   Personality

Personality is that pattern of characteristic thoughts, feelings, and behaviours that distinguishes one person from another and that persists over time and situations [6]. Personality is an important factor in human behaviour since humans with the same goal and in similar circumstances may behave differently according to their personalities. In our context, two drivers may behave differently under the same circumstances.

In [7], it is stated that personality traits and gender were found to explain 37.3% of the variance in risky driving behaviour. In [8] a study to understand the relationship between personality and the number of fines received stated that conscientiousness factor was a key to negatively predict the number and amount of financial fines the drivers had during the last three years. The openness factor positively predicted the number of fines they had in the last 3 years and the amount of financial fines during the last year. The extraversion factor both meaningfully and positively could predict only the amount of financial fines they had during the last year [8]. According to [9] Personality and driving behaviour have strong correlations [10]. Again, [11] is quoted in [9] to claim that most studies found significant positive relations between sensation seeking and aspects of aggressive and risky driving. The studies mentioned above all point to the same intuitive fact, namely that the personality is a key influencing factor in drivers' behaviour.

There are many personality models that consist of a set of dimensions, where every dimension is a specific property [12]. In the past two decades there has been remarkable progress in one of the oldest branches of personality psychology: the study of traits or individual differences [13]. To integrate the personality factor in our modelling we need to select a personality model. In [13] there is a growing agreement among personality psychologists that most individual differences in personality can be understood in terms of five basic dimensions: Neuroticism (N) vs. Emotional Stability; Extraversion (E) or Surgency; Openness to Experience (O) or Intellect; Agreeableness (A) vs. Antagonism; and Conscientiousness (C) or Will to Achieve [14], [15], [16].

In [13] the five factors are described as follows:

- *Neuroticism*, it represents the individual's tendency to experience psychological distress, and high standing on N is a feature of most psychiatric conditions. Indeed, differential diagnosis often amounts to a determination of which aspect of N (e.g., anxiety or depression) is most prominent.

- *Extraversion* is the dimension underlying a broad group of traits; including sociability, activity, and the tendency to experience positive emotions such as joy and pleasure. Patients with histrionic and schizoid personality disorders differ primarily along this dimension. [17] and [18] have pointed out that talkative extraverts respond very differently to talk-oriented psychotherapies than do reserved and reticent introverts.

- *Openness* is how ready you are to Experience. High-O individuals are imaginative and sensitive to art and beauty and have a rich and complex emotional life. They are intellectually curious, behaviourally flexible, and nondogmatic in their attitudes and values [19].

- *Agreeableness*, like E, is primarily a dimension of interpersonal behaviour. High-A individuals are trusting, sympathetic, and cooperative; low-A individuals are cynical, callous, and antagonistic.

- *Conscientiousness* is a dimension that contrasts scrupulous, well-organized, and diligent people with lax, disorganized, and lackadaisical individuals. Conscientiousness is associated with academic and vocational success [20].

Each dimension has six facets [12]. For the sake of simplicity these facets are not covered in this research.

Several tests exist to measure the personality traits and their respective facets or just the traits. Among the existing tests we can mention The NEO PI-R and The NEO-FFI. According to [21] NEO PI-R is a 240-item inventory developed by Paul Costa and Jeff McCrae. It measures not only the Big Five (the five traits), but also six "facets" (subordinate dimensions) of each of the Big Five [21]. The NEO PI-R is a commercial product, controlled by a for-profit corporation that expects people to get permission and, in many cases; pay to use it [21]. Costa and McCrae have also created the NEO-FFI, a 60-item truncated version of the NEO PI-R that only measures the five factors. The NEO-FFI is also commercially controlled [21].

Another alternative test is The International Personality Item Pool. According to [21] it is developed and maintained by Lew Goldberg, has scales constructed to work as analogs to the commercial NEO PI-R and NEO-FFI scales (see below). IPIP scales are 100% public domain - no permission is required for their use.

We choose the OCEAN model because it is the most accepted model among psychologists and the availability of free questionnaire to measure the personality traits according to it.

3.2   Emotions

According to [22] emotions were seen as an undesirable product of the human mind. Therefore the less the person is emotional the more intelligent and desirable he/she is. An opposite view says that currently, researchers claim that emotions are part of life and are necessary in intelligent behaviour [23], [24]. We outline the importance of emotions in driving behaviour in what follows.

3.2.1   Importance of Emotions in Driving Behaviour

In [25] an experience has been done with happy, sad and neutral music alternated with no-music phases while driving in a simulator. Results showed that happy music distracted drivers as their mean speed unexpectedly decreased and their lateral control deteriorated. Sad music caused drivers to drive slowly and kept their vehicle in its lane.

According to [25] in one study, the author noticed that drivers who experienced anger accelerated and committed more traffic violations than others and [26] concluded that emotions influence traffic risk evaluation and general driving behaviour. [25] quote the work of [27] and [28] stating that anger is one of the most common negative emotions experienced during driving, leading to aggressive driving behaviour. In this sense, angry drivers intentionally endanger others with aggressive verbal and/or physical expressions [25]. Emotions like sadness, discontentment or joy are likely to impact on attention, leading to a different driving style. According to [29] a research study looked into the cause of accidents. Driver behaviour is one of the main reasons for this predicament while emotion plays a vital role as it affects the driver's behaviour itself. In understanding the correlation between drivers' behaviour and emotion, the analysis results of an experience conducted by [29] showed that for each driver pre- cursor emotion will affect the emotion in pre-accident whereas negative emotions appear frequently in post-accident compared to positive emotion. The studies above seem to be pointing to the fact that emotions have a direct impact on drivers' behaviour. For this reason a model for emotions is considered as a key component of our system.

3.2.2   Emotion Models

Most evaluation models are based on the principle that emotions arise as a consequence to a cognitive evaluation of the environment [30]. According to [30], Lazarus [31] created a model of assessment in which he unified evaluation (appraisal) and adaptation (coping). In fact, he distinguished between two types of evaluation: (1) primary, which assesses the relevance of an event and its congruence with the goals or not, (2) the secondary assessing what can or should be done to respond to this event [30]. According to [30], Roseman and Spindel [32] created a model in which he identified five criteria for evaluating events. Depending on the values of these criteria they characterize thirteen distinct emotions. The first criterion determines whether a situation is positive or negative relative to the goals of the individual. The second criterion determines whether the situation is in agreement or not with the state of motivation. The third criterion is related to the certainty or uncertainty of the event. The fourth criterion defines whether a person perceives himself in a given situation, as strong or weak. The fifth criterion is the origin of the event, whether it is related to the circumstances, or rather linked to the individual himself or others. Among these models the OCC model [33], is one of the most used in computing [30]. In OCC the authors define 22 types of emotion. Same types of emotions are triggered in similar situations. To every emotion is defined an opposite emotion. For example, sadness is the opposite of joy [30]. More recently, Ortony (one of the authors of the model) has simplified OCC model by grouping types of emotions to finally define five types of positive emotions and 6 types of negative emotions [34]. According to [34] in the OCC model, three classes are defined according to their triggering cause:

- The emotions triggered by an event affecting an object of the individual, such as joy, hope or fear.
- The emotions triggered by an event affecting a principle or standard, such as shame, pride or reproach.
- The emotions triggered by the perception of particular objects (animated or unanimated, concrete or abstract) such as love or hate.

The authors define global variables that determine the intensity of the emotions and local variables (specific to each of the above classes) that will determine both the type and intensity of the emotions of each class. This model is the most widely used model to simulate processes triggering emotions in computational systems [34]. The figure 1 represents the original structure of the OCC model.

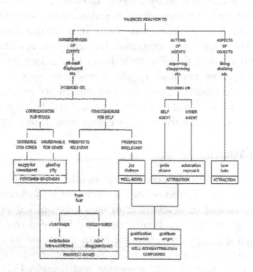

Fig. 1 Original structure of the OCC model [33]

Let's assume that the person being considered (as a potential case for road rage) is described by a software agent. Let us also define the agent's external world as a set of events on the agent, actions of other agents and the agent's own perceptions of objects. Similarly, we define the agent's internal world as a set of agent's goals, its standards and attitudes. These definitions are based on the work by [33]. These two worlds will interact and trigger any one or more of the 22 emotions. The concrete definition of goals, standards and attitudes are domain specific and should be defined by the final user. Figure 2 shows the specifications of the 22 emotions presented in the OCC model.

The emotional model we propose in the next section is based on the OCC model due to its simplicity, but also to the fairly comprehensive types of emotions covered.

```
              Joy: (pleased about) a desirable event
          Distress: (displeased about) an undesirable event
       Happy-for: (pleased about) an event presumed to be desirable for someone else
            Pity: (displeased about) an event presumed to be undesirable for someone else
          Gloating: (pleased about) an event presumed to be undesirable for someone else
    Resentment: (displeased about) an event presumed to be desirable for someone else
            Hope: (pleased about) the prospect of a desirable event
            Fear: (displeased about) the prospect of an undesirable event
      Satisfaction: (pleased about) the confirmation of the prospect of a desirable event
Fears-confirmed: (displeased about) the confirmation of the prospect of an undesirable event
          Relief: (pleased about) the disconfirmation of the prospect of an undesirable event
  Disappointment: (displeased about) the disconfirmation of the prospect of a desirable event
            Pride: (approving of) one's own praiseworthy action
          Shame: (disapproving of) one's own blameworthy action
        Admiration: (approving of) someone else's praiseworthy action
        Reproach: (disapproving of) someone else's blameworthy action
     Gratification: (approving of) one's own praiseworthy action and (being pleased about) the related desirable event
        Remorse: (disapproving of) one's own blameworthy action and (being displeased about) the related undesirable event
        Gratitude: (approving of) someone else's praiseworthy action and (being pleased about) the related desirable event
            Anger: (disapproving of) someone else's blameworthy action and
                   (being displeased about) the related undesirable event
            Love: (liking) an appealing object
            Hate: (disliking) an unappealing object
```

Fig. 2 Specifications of the 22 types of emotions [33]

3.3   Social and Personal Attributes

In what follows, we present the effects of personal/social attributes of drivers on their driving behaviour. We included most of these attributes in our model. These are presented below. For the few that are not included, reasons of their omission are given.

- *Gender*: In [7], it is stated that personality traits and gender were found to explain 37.3% of the variance in risky driving behaviour.
- Age: A survey done by Jonah [11] states that young drivers [16, 24] are more likely to engage in dangerous driving. [16, 19] age group were more likely to have more accidents and violation rates than the others groups.
- *Knowledge of the area*: According to [9] Route knowledge has been identified as important for the driving task [36].
- *Fatigue*: Many papers stated that fatigue has an influence on driver's perceptions and evaluation/ response time.
- *Type of the car*: Depending on the type of the car some people will drive differently. For example a driver using a Land Rover will drive with a different way than driving a small car. On the other hand some studies state that an over reliance on automation will decrease driver's vigilance.

Based on the findings of the studies above, the following parameters will be incorporated in our model: Gender, Age, Knowledge of the area, Type of the car. We don't include fatigue for the sake of simplicity of the modelling.

3.4 Cognitive Reactions

To every situation an agent could have a different reaction. This reaction could depend on his personality, emotions or culture. The reactions of the agent (driver) could have different impacts on the future outcomes of an encounter, e.g.: an encounter might be a conflict between one driver and another. Depending on the reactions we may boost the actual rage or lower it. For example, if a driver A is blocking driver B and driver A is driving at a low speed. If B replies to A with an insult or a gesture, A might do an action frustrating B even more. On the other hand if driver B uses less aggressive manners to express his anger, A's reaction may be less aggressive too. Such actions may depend on the personality and the emotions but they may also depend on the values of the persons. Is insulting someone OK for our self-image or is it shameful? It is the society's interpretation of the possible reactions and to what extent a driver would go in a given situation. One of the solutions we will exploit to alleviate the road rage problem is influencing the reactions of people in negative emotional states. A way to influence those reactions is by influencing the values of the society.

4   Proposed Emotional Model

Our agents will incorporate an emotional model. This model is used to simulate human emotions by the agent. The agents are tasked to simulate the entire driver behaviour. The emotional module is a part of this entire behaviour.

4.1 Architecture of the Model

Our emotional model will be based on the OCC model. Its aim is to predict the emotions felt by the drivers and their intensities. The OCC model defines the standards, goals and attitudes [33] to evaluate the intensities of the felt emotions in any particular group (event, object or agent) alongside with other variables. We add to our model "the beliefs". The beliefs are what a driver believe about a current situation and how does it interact with his standards, goals and attitudes. We have chosen to use a probabilistic form of cognitive maps to represent beliefs and their co-influence. In real life, a belief can influence another belief, e.g. if I believe that if it rains then a football game may be delayed. The word "may" can be translated into a probability. Thus, the belief "It rains" and the belief "Football game delayed" are in a relationship of effect with a probability. Therefore, we believe that a probabilistic cognitive map is a good representation of the beliefs and their relationships.

We distinguish two types of nodes in our cognitive map. The first type are nodes representing a belief. The second type are nodes representing an effect. The effect nodes will not have any incoming arcs (input) and they are marked with the prefix "Eff". The belief nodes are nodes representing a possible belief in the current situation. The weights of the "arcs" are the probability of the belief to be true. For example if we have an "arc" from node A to Node B weighted to X, this means that if A is true then B has a probability of X to be true.

In our model, emotions are triggered by modifications of beliefs and not directly via events, actions of agents or objects. An effect could trigger an event, an action of an agent or the visibility of an object. Such unification of events, action of agents and objects in effects is due to the fact that any situation could create many events, actions or objects perceived at the same time. As an example: consider a person 'A' driving a car who accidentally crashes into a driver 'B'. Driver 'A' may focus on the damage of the car and then perceive what happened as an event and feel distress. Driver 'A' may focus on driver 'B' as a driver who is causing damage to him (as an agent) and then feels reproach. And the last scenario, is that driver 'A' will perceive driver 'B' as a reckless driver (object) and feel hate for him. According to [33] some of those emotions could occur (e.g. 'A' feels only distress) and some could occur successively; depending on the person experiencing the situation.

We believe that an effect could trigger many emotions at the same time and then cause the change of several beliefs. An effect may not trigger the same emotions in everybody. It could produce many events, actions and perceived objects but this production could be different from one person to another. The order of perception may be different from one person to another too. An example of the probabilistic cognitive map is presented in figure 3.

To determine which emotion is going to be felt after the modification of a belief, we trace this modification to the effect triggering it. If this effect is perceived as an event we will trigger an event-based emotion. The desirability of the variation in the belief and its effects on the goals/standards/attitudes will determine whether the emotion is positive or negative. Some beliefs' variations might not trigger any emotions directly.

We plan to use the profile (personality + social parameters+ previous emotions) of the person to predict what a given driver may perceive and in what order through live experience with drivers.

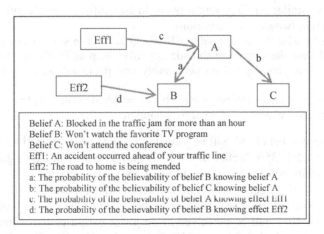

Fig. 3 An example of the probabilistic cognitive map

4.2   Computation of the Intensities

Once the design of the probabilistic cognitive map completed, we proceed with defining the intensities of the emotions felt. The intensities will depend on the beliefs variations, the personality, the importance of the goals/standards/attitudes interacting with the targeted belief (e.g. in completing the goals positively or negatively), the surprise element and the intensity of the current emotions.

The intensity of the felt emotions will decrease over time. The decreasing speed will depend on the personality and social/personal attributes. Prospect relevant emotions like fear and hope will not decrease over time but will turn into relieved/fear confirmed and satisfaction/disappointment respectively depending on the confirmation of those two emotions.

4.3   Scenarios

Here are scenarios giving more explanations about the emotional model:

Let a person named Bob who wants to go to a stadium to watch a football game. Figure 4 presents the cognitive map of Bob's beliefs. Here is the description of those beliefs:

A: Traffic jam problem
B: Will be late to the stadium
C: The car was damaged (Event)
D: Seeing an unappealing object
E: A driver has damaged the car (action of an agent)

Eff1 and Eff2 are respectively an official visiting the town and a driver crashing into Bob's car causing damage to the car. As we can see the beliefs C and E are two different beliefs. These beliefs are triggered by the same effect which is "a driver crashing into Bob's car causing damage to your car" and although they appear to be similar but they might occur in two different consecutive moments and they trigger two different emotions.

Any driver with the same goals as Bob's goals and in similar external circumstances will have the same probabilistic cognitive map as Bob's but with different weights. The weights depend on personality and personal/social attributes of the driver.

Bob's weights are as follows: a=0.7, b= 1, c= 0.5, d= 0.8, e=1, f=1
Here we have two examples of what may happen:

- If Eff1 occurs, belief 'A' will be true at 70% and belief 'B' will be true at 70% X 50% which is 35%. Belief 'B' will have an influence on Bob's goal (arriving to the stadium before the starting of the game). The variation of the veracity of belief 'B' will trigger an emotion. To detect which emotion will be triggered we need to look at the effect that was at the origin of the variation of belief 'B'. In this case Eff1 will be perceived as an event with a negative impact on the driver. Because belief 'B' is not 100% true, the emotions that will be felt are prospect relevant. In this case Bob will feel 'Fear'. The intensity of this emotion will be determined by the personality/social attributes, surprise, the previous emotional state and the importance of the goal. If the goal is met 'Bob' will feel relieved otherwise he will have his fears confirmed. In our model prospect based emotions will not decrease over time because this type of emotion depends on a probability and not on a fact; a belief not true yet but may be true in the future. After their confirmation or disconfirmation the re-sulted emotion will decrease over time. Here the emotion *relief* and *fears-confirmed* will decrease over time (if felt).

- If 'Eff2' occurs, Bob may perceive this effect from three different angles. He will perceive it as an event, an action of an agent and an object. Bob's order of perception will not be discussed here. If he starts by perceiving 'Eff2' as an event then belief 'C' will be true at 90%. Belief 'B' will be affected also and will be true at 100% X 80% which is 80%. The variation of belief 'B' will in teract with a goal and is originally triggered by an event. Thus, 'Bob' will per-ceive *fear*. The intensity will depend on the same parameters cited above.

'Bob' will then perceive 'Eff2' as an action of an agent. Belief 'E' will be true at 100%. This belief has an interaction with the standards of 'Bob', which will trigger action based emotions. In this case it is '*reproach*' that will be felt. The intensity depends on the same parameters mentioned above. The '*reproach*' feeling will decrease over time.

Finally 'Bob' will perceive 'Eff2' as an unappealing object which will trigger '*hate*'.

As mentioned above the order in which these perceptions (event, action and ob-ject) occur will not be discussed here.

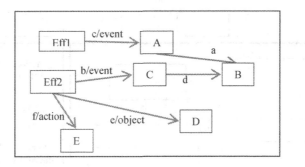

Fig. 4 Bob's probabilistic cognitive map

5   Modelling the Behaviour

After modelling the emotional module, the next stage will be modelling the behav-iour of the drivers. The behaviour of the driver will depend on the goals, personal-ity, emotions, social/personal attributes and the values (a set of expected cognitive reactions) of the driver. The emotional module will act as a precursor to the behav-ioural module. A classification method must be defined (neural networks, genetic algorithms, neuro-fuzzy) to predict drivers' behaviours using the selected parameters. The classes that the classification method will have to predict are the possible reactions.

After modelling the entire drivers' behaviours, we will simulate a whole driving behaviours in a virtual city with different types of drivers with different goals for each. Each driver is represented by an agent. Agents are composed of an emotional module and a behavioural module. Emotional module is meant to compute agent's emotions and their intensities. Behavioural module will use all the parameters above to produce agent's behaviour. The environment of the agents is a representation of a 3D city and its roads. In this environment, there will be a lot of interactions between agents since every agent has the possibility to initiate many effects and then triggering a set of emotions on the other agents. The figure 5 presents the structure of agents.

We will use a monitoring system to observe the development of road rage and the actions leading to road rage. We will try some possible solutions in our system (like influencing personalities, influencing the values…) and evaluate them. The solutions we will propose will not concern driver's capabilities as we assume that most drivers have a minimum level of driving aptitude. Our focus will be on psychological and social aspects.

A 3D visualisation module is developed to monitor in a 3D environment the dynamics of the city.

We plan to compute the number of negative actions carried out by the drivers and how the propagation of those actions will affect them in the future. The aim of this is to make the drivers aware of their actions and the impact of their actions on them and other users of the road.

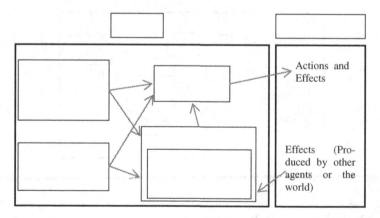

Fig. 5 Structure of agent

6   Conclusion and Future Work

Modelling human's behaviour is not an easy task. Its complexity is as complex as humans are. We needed to look at what constitute human's cognition. The number of variables influencing human's behaviour is very large. However, when looking at it in the context of a specific aspect like driving behaviour, we could select less parameters than the parameters needed to model an entire human's behaviour. Those selected parameters do not have the same importance during modelling the different stages that constitute driver's behaviour.

In this study, we started by exploring what influences drivers' behaviour. For this we selected personality, emotions, social/personal parameters and values. For *personality* we used the OCEAN model and for *emotion* we proposed our own model based on the OCC model where we have defined a set of interactions influencing emotions. In the future, we will specify the interactions between various types of simultaneous emotions and their impact on each other. For the sake of simplicity we did not include driving abilities. Instead, we assume that all of the drivers have a minimum level of driving skills. The solutions that we will be proposing are psychological/social rather than based on driving skills.

The next stage in this research is to model the entire driving behaviour using machine learning techniques.

References

1. Takaku, S.: Reducing road rage: An application of the dissonance-attribution model of interpersonal forgiveness. Journal of Applied Social Psychology 36, 2362–2378 (2006)
2. National Highway Traffic Safety Administration: Welcome to Stop Aggressive Driving (2004), http://www.nhtsa.dot.gov/people/injury/aggressive/Aggressive%20Web/sse_1.html (accessed November 20, 2012)
3. American Automobile Association Foundation for Traffic Safety. Aggressive driving: Three studies. Author, Washington, DC (1997)
4. James, L., Nahl, D.: Road rage and aggressive driving: Steering clear of highway warfare, Amherst, Prometheus (2000)
5. Sharkin, B.S.: Road rage: Risk factors, assessment, and intervention strategies. Journal of Counseling and Development 82, 191–198 (2004)
6. Phares, E.J.: Introduction to Psychology, 3rd edn. Harper Collins Publishers, New York (1991)
7. Oltedal, S., Rundmo, T.: The effects of personality and gender on risky driving behaviour and accident involvement. Safety Science 44, 621–628 (2006)
8. Esmaeili, B., Far, H.R.I., Hosseini, H., Sharifi, M.: The Relationship between Personality Characteristics and Driving Behavior. World Academy of Science, Engineering and Technology 67 (2012)
9. Oppenheim, I., Shinar, D., Carsten, O., Barnard, Y., Lai, F., Vanderhaegen, F., Polet, P., Enjalbert, S., Pichon, M., Hasewinkel, H., Lützhöft, M., Kircher, A., Kecklund, L.: Critical review of models and parameters for Driver models in different surface transport systems and in different safety critical situations. ITERATE Delivrable 1.1 (2010)
10. Sumer, N., Lajunen, T., Ozkan, T.: Big Five personality traits as the distal predictors of road accident involvement. In: Underwood, G. (ed.) Traffic and Transport Psychology, pp. 215–227. Elsevier, Oxford (2005)
11. Jonah, B.A.: Sensation seeking and risky driving: A review and synthesis of the literature. Accident Analysis and Prevention 29, 651–665 (1997)

12. Ghasem-Aghaee, N., Khalesi, B., Kazemifard, M., Ören, T.I.: Anger and Aggressive Behavior in Agent Simulation. In: Summer Computer Simulation Conference, Istanbul, Turkey, pp. 267–274 (2009)
13. Costa Jr., P.T., McCrae, R.R.: NEO PI-R Professional Manual. Psychological Assessment Resources, Inc., Odessa (1992)
14. Digman, J.M.: Personality structure: Emergence of the five-factor model. Annual Review of Psychology 41, 417–440 (1990)
15. John, O.P.: The "big five" factor taxonomy: Dimensions of personality in the natural language and in questionnaires. In: Pervin, L. (ed.) Handbook of Personality Theory and Research, pp. 66–100. Guilford, New York (1990)
16. Norman, W.X.: Toward an adequate taxonomy of personality attributes: Replicated factor structure in peer nomination personality ratings. Journal of Abnormal and Social Psychology 66, 574–583 (1963)
17. Wiggins, J.S., Pincus, A.L.: Conceptions of personality disorders and dimensions of personality. Psychological Assessment: A Journal of Consulting and Clinical Psychology 1, 305–316 (1989)
18. Miller, X.: The psychotherapeutic utility of the five-factor model of personality: A clinician's experience. Journal of Personality Assessment 57, 415–433 (1991)
19. McCrae, R.R., Costa Jr., P.T.: Openness to experience. In: Hogan, R., Jones, W.H. (eds.) Perspectives in Personality, vol. 1, pp. 145–172. JAI Press, Greenwich (1985)
20. Digman, J.M., Xakemoto-Chock, N.K.: Factors in the natural language of personality: Re-analysis, comparison, and interpretation of six major studies. Multivariate Behavioral Research 16, 149–170 (1981)
21. Srivastava, S.: Measuring the Big Five Personality Factors (2012), http://psdlab.uoregon.edu/bigfive.html (accessed November 29, 2012)
22. Miranda, J.M.: Modelling human behaviour at work: an agent-based simulation to support the configuration of work teams. PhD thesis, Universidad complutense de Madrid, Madrid (2011)
23. Damasio, A.R.: Descartes' Error: Emotion, Reason and the Human Brain. Grosset/Putnam, New York (1994)
24. Gray, J., Braver, T., Raichele, M.: Integration of Emotion and Cognition in the Lateral Prefrontal Cortex. National Academy of Sciences USA, 4115–4120 (2002)
25. Pêcher, C., Lemercier, C., Cellier, J.M.: Emotions drive attention: Effects on driver's behaviour. Safety Science 47(9), 1254–1259 (2009)
26. Mesken, J.: Determinants and consequence of drivers' emotions. Ph.D. thesis, University of Groningen and SWOV, Groningen (2006)
27. Deffenbacher, J.L., Lynch, R.S., Oetting, E.R., Swaim, R.C.: The Driving Anger Expression Inventory: a measure of how people express their anger on the road. Behaviour Research and Therapy 40, 717–737 (2002)
28. Mesken, J., Hagenzieker, M.P., Rothengatter, T., de Waard, D.: Frequency, determinants, and consequences of different drivers' emotions: an on-the-road study using self-reports (observed) behaviour, and physiology. Transportation Research Part F 10, 458–475 (2007)
29. Nor, N.M., Wahab, A., Kamaruddin, N., Majid, M.: Pre-post accident analysis relates to pre-cursor emotion for driver behavior understanding. In: The 11th WSEAS International Conference on Applied Computer Science (ACS 2011), pp. 152–157 (2011)

30. Chaffar, S., Frasson, C.: Apprentissage machine pour la prédiction de la réaction émotionnelle de l'apprenant. Revue des Sciences et Technologies de l'Information et de la Communication pour l'Éducation et la Formation (STICEF), numéro spécial sur les dimensions émotionnelles de l'interaction en EIAH, 14 (2007)
31. Lazarus, R.: Emotion and Adaptation. Oxford University Press, NY (1991)
32. Roseman, I.J., Jose, P.E., Spindel, M.S.: Appraisals of Emotion-Eliciting Events: Testing a Theory of Discrete Emotions. Journal of Personality and Social Psychology 59(5), 899–915 (1990)
33. Ortony, A., Clore, G., Collins, A.: The Cognitive Structure of Emotions. University Press, Cambridge (1988)
34. Ochs, M.: Modélisation, formalisation et mise en œuvre d'un agent rationnel dialoguant émotionnel empathique. Ph.D Thesis, Université Paris VIII, Paris (2007)
35. Jonah, B.A.: Age differences in risky driving. Health Education Research. Theory & Practice 5(2), 139–149 (1990)
36. Luther, R., Livingstone, H., Gipson, T., Grimes, E.: Understanding driver route knowledge. In: Wilson, J.R., Norris, B., Clarke, T., Mills, A. (eds.) People and Rail Systems: Human Factors at the Heart of the Railway, pp. 71–78. Ashgate, Aldershot (2007)

Making Existing Reactive Systems Anticipatory

Kai Shi, Yuichi Goto, Zhiliang Zhu, and Jingde Cheng

Abstract. From the viewpoints of high safety and high security, any critical reactive system should be anticipatory, i.e., the system should be able to detect and predict accidents/attacks, take some actions to inform its users, and perform some operations to defend the system from possible accidents/attacks anticipatorily. However, most of existing reactive systems are not so, furthermore, it is impractical, but not impossible, to rebuild them to be anticipatory, because reimplementation of the whole of a system results in high cost. Therefore, it is desirable to extend an existing legacy reactive system with anticipatory ability without reimplementing the whole of the legacy system. This paper proposes a general methodology to realize such an extension. The novelty of the methodology is that it does not require reimplementation the whole legacy system, does not affect the system's original functions, and can deal with various reactive systems by using the same process. By extending an existing reactive system anticipatory using our methodology, we can get a new generation system with high safety and high security. This paper also presents three case studies to show how to apply our methodology.

1 Introduction

A reactive system maintains an ongoing interaction with its environment, as opposed to obtain a final result [10]. Various reactive systems play very important roles in modern society, such as bank transfer systems, web servers, operating systems, computer networks, air/railway traffic control systems, elevator systems, and nuclear power plant control systems. Since an accident or an attack of a critical

Kai Shi · Yuichi Goto · Jingde Cheng
Department of Information and Computer Sciences,
Saitama University, Saitama, 338-8570, Japan
e-mail: {shikai,gotoh,cheng}@aise.ics.saitama-u.ac.jp

Kai Shi · Zhiliang Zhu
Software College, Northeastern University, Shenyang, 110819, China
e-mail: {shik,zhuzl}@swc.neu.edu.cn

R. Lee (Ed.): *Computer and Information Science*, SCI 493, pp. 17–32.
DOI: 10.1007/978-3-319-00804-2_2   © Springer International Publishing Switzerland 2013

reactive system may cause financial loss and even casualties, the biggest challenge for reactive systems is not only to ensure the system functionality, but to prevent these accidents and attacks [1, 14].

A traditional reactive system is passive, i.e., it only performs those operations in response to instructions explicitly issued by users or application programs, but have no ability to do something actively and anticipatorily by itself. Therefore, a passive reactive system only has some quite weak capability to defend accidents and attacks from its external environment. In order to prevent accidents/attacks beforehand, it is desired that a reactive system is *anticipatory*, i.e., the system should be able to detect and predict accidents/attacks, take some actions to inform its users, and perform some operations to defend the system from possible accidents/attacks anticipatorily. From the viewpoints of high safety and high security, any critical reactive system should be anticipatory [3].

To build practical anticipatory systems [20] with high safety and high security, Cheng proposed *anticipatory reasoning-reacting system* (ARRS) [3], which is a computing system that can predict based on the predictive model, then take anticipatory actions according to the predictions as well as take reactive actions to the current situation based on the behavioral model. In other word, an ARRS is a reactive system with ability of anticipation. Some prototypes of ARRSs were implemented [12, 22].

The problem is that most of existing reactive systems are not anticipatory, furthermore, it is impractical, but not impossible, to rebuild them to be anticipatory, because reimplementation of the whole of a system results in high cost. If we can extend an existing reactive system with anticipatory ability but preserving the system's original functions, we can improve the system safety or security at a relatively small cost. However, there is no study about whether it is possible to extend an existing reactive system to be an anticipatory and how to realize such an extension.

This paper argues it is possible to extend an existing legacy reactive system with anticipatory ability without reimplementing the whole of the legacy system, and proposes a general methodology to realize such an extension. By extending an existing reactive system anticipatory using our methodology, we can get a new generation system, indeed an ARRS, with higher safety and higher security. This paper also presents three case studies to show how to apply our methodology to legacy reactive systems.

The contributions of this paper are: 1) to propose a methodology which can extend various reactive systems with anticipatory ability as well as preserving the system's original functions without reimplementing the whole of the legacy system, and 2) through case studies to show the usefulness of anticipatory reasoning-reacting systems.

2 Feasibility Analysis

Any reactive system M has a set I of inputs (e.g., inputs that represent events and conditions produced by the environment), a set O of outputs (e.g., action instructions produced by M), and a reactive function f_P mapping non-empty sequences of subset of I into a subset of O [10, 17]. Many reactive systems also have some implemented mechanisms to take action instructions to prevent an ongoing/potential accident/attack, such as applying the brake in a railway system, and blocking an intruder by configuring a firewall in computer networks. However, a traditional reactive system cannot take these action instructions anticipatorily, because f_P is passive. Therefore, we can add an additional anticipatory function f_A shown in figure 1, which can detect and predict accidents/attacks base on I, then choose appropriate action instructions to handle these accidents/attacks, which can be taken by the original reactive system. For most concurrent systems, the f_A only carries out some additional action instructions, while the system's original reactive function is kept, which means the original reactive actions and the anticipatory actions are taken simultaneously. For those systems that anticipatory actions may affect the system's original reactive actions such as a critical sequence of actions, in most cases, the system's original reactive actions have a higher priority thus the anticipatory actions must wait, because we do not want to affect the system's original functions. However, in some situation, the original reactive actions may make against the safety or security of the system, the original actions should be canceled, while the system's original functions are still kept in most of the time. We call the original reactive system *legacy system*, and call the extended system *target system*, as the figure 1 shows.

As we discussed above, if we want to extend a legacy reactive system with anticipation, the system must satisfy the following requirements.

- There must be some approaches to perceive adequate subset of I of the legacy system for detection and prediction.
- The legacy system must have some implemented mechanism to take actions to handle the ongoing/potential accidents/attacks.

Besides, for some extreme real time systems, it may be too late to predict and take actions before the accidents/attacks cause pernicious consequences, thus such systems are not suitable to extend with anticipation.

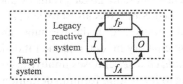

Fig. 1 Extending a legacy reactive system with anticipatory function

ARRSs are suitable candidates of target systems for the extension. The most important features of ARRSs are: 1) both the prediction and decision-making are base on logic-based forward reasoning, and 2) an ARRS is an extension of a reactive system. Logic reasoning based prediction and decision-making provides an area-independent, task-independent, general-purpose reasoning mechanism [3, 6], which can apply to different application area of the legacy systems. As an extension of a reactive system, the ARRS's architecture [9, 22] is fit to embrace an existing reactive system.

3 Methodology

3.1 Target System

Our methodology aims to combine an existing reactive system, called *legacy system*, with some components which can provide anticipatory ability, then get a new system that in fact is an ARRS, called *target system*, which can deal with accidents/attacks anticipatorily with preserving the original system functions. Figure 2 shows a general architecture of the target system embracing a legacy reactive system. Figure 3 shows the data flow diagram of the target system. Besides the legacy system, the target system includes following components.

There are three databases. *LTDB* is a logical theorem database, which stores fragments of logical systems [3]. *ETDB* is an empirical theory database storing anticipatory model. *RDB* is a rule database storing *filter rules*, *translation rules*, *interesting formula definitions*, *interesting terms*, *actions mapping rules*, and *condition-action rules*.

Observers and filter deal with the data from the legacy system. *Observers* perceive the inputs of the legacy system that represent events and conditions produced by the environment. Besides, the observers could also perceive the status of the legacy system. In many cases, there are some sensors/monitors/statistical tools have been implemented in the legacy system. If we can utilize these observers, we do not need to implement new observers. *Filter* filters out the trivial sensory data and generates important and useful information for detection/prediction or decision-making.

Formula generator, forward reasoning engine, and formula chooser generate predictions or next actions. *Formula generator* encodes the sensory data into logical formulas used by the forward reasoning engine according to the translation rules. *Forward reasoning engine* is a program to automatically draw new conclusions by repeatedly applying inference rules, to given premises and obtained conclusions until some previously specified conditions are satisfied [7]. The forward reasoning engine gets logical formulas translated at the formula generator, fragment of (a) logic system(s), and the anticipatory model, and then it deduces candidates of predictions/next actions. *Formula chooser* chooses nontrivial conclusions from forward reasoning engine according to interesting formula definitions and interesting terms.

Action planner receives candidates of next actions and current situation, and then calculate the planned actions. The action planner has two functions: 1) the candidates of actions are only based on qualitative information of the situation, while the action planner can utilize quantitative information to revise the actions, and 2) to make accurate plan, calculation is needed, such as to calculate how many special actions should be taken to prevent a certain accident/attack.

Enactor receives planned actions from action planner, matches the current situations to the situation-action rules to select appropriate actions to take, and then gives corresponding instructions according to the actions mapping rules to invoke the actions of the legacy system. Besides, if the original reactive actions of the legacy system may make against the safety or security of the system in some situation, the enactor are also in charge of canceling these original actions. Condition-action rules specify when an unexpected accident/attack occurs abruptly in real time, 1) which reactive actions should be taken, and 2) which planned actions should be canceled.

Soft system bus [4] and *central control components* provide a communication channel for each component and an infrastructure for persistent computing [4].

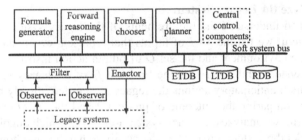

Fig. 2 Architecture of the target system

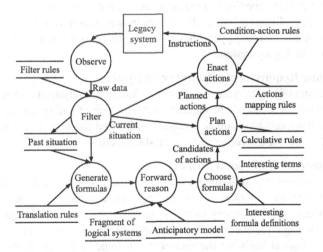

Fig. 3 Data flow diagram of the target system

3.2 Phases

In our methodology, there are five major phases to extend a legacy reactive system with anticipatory ability.

Phase 1: Analyze the Target Domain
First, we analyze the possible accidents/attacks in the legacy system's target domain. We list possible accidents/attacks in the target domain, and then for each possible accident/attack. We analyze each possible accident's/attack's causation, formation process, and the consequence, as well as other attributes such as likelihood and severity. Based on the severity and likelihood, we could sort these possible accidents/attacks in order of importance.

Second, we analyze the possible (anticipatory) actions for defending against accidents/attacks in the target domain. For each possible accident/attack, we find out whether it is possible to defend against it and what actions could defend against it, both an ongoing one and a potential one.

Phase 2: Analyze the Legacy System
First, we need to understand the requirements of the legacy system. Second, for the legacy system, we list the set I of inputs that represent events and conditions produced by the environment and the set O of output action instructions produced by the legacy system, and understand the function f_P of the legacy system. Third, we find out which anticipatory actions the legacy system has already had the ability to take, by comparing the outcome of phase 1.2 and the set O of outcome of phase 2.2. Fourth, we analyze the legacy system to find out which ongoing/potential accidents/attacks the system already has the ability to handle, and how the system defends against them. After that, we could also find out which accidents/attacks that the legacy system cannot handle or do not handle well, which may be the goals of the target system. Fifth, we find out the approaches to perceive the inputs that represent events and conditions produced by the environment and (optional) status of the legacy system itself, and the approaches to invoke these existing anticipatory action instructions of the legacy system.

Phase 3: Define Requirements of the Target System
The requirements of target system include the functional requirements that specify which accidents/attacks should be tackled, and which anticipatory actions should be used to prevent against a certain accidents/attacks, and the non-functional requirements including performance, security, availability, etc.

Phase 4: Construct Anticipatory Model
First, before constructing anticipatory model, we need to choose the logic basic(s) for the target domain. A model is a simplified representation of something [8]. In the target system, an anticipatory model is represented by a set of logical formulas of a specific logic (or logics). Thus before we build an anticipatory model, we must choose a logical basis for the model. Such logic must satisfy the following requirements [3]. First, the logic must be able to underlie relevant reasoning as well as truth-preserving reasoning in the sense of conditional. Second, the logic must be able to underlie ampliative reasoning. Third, the logic must be able to underlie para-complete and paraconsistent reasoning. Especially, the logic for prediction must be able to underlie temporal reasoning.

Second, we construct world model. A world model represents status of the real world and essential empirical knowledge (not related with time and behavior) in target domain. We use a vocabulary of predicates or constant terms to present the status of the real world. In addition, if necessary, we collect and represent empirical knowledge (not related with time and behavior) as conditionals (see "conditional 2" [8]). The construction of the world model involves following steps. 1) In the target domain, list possible objects and their properties and statuses. 2) List relationships among the objects (if have). 3) List possible events and conditions produced by the environment we concern. 4) Determine the essential empirical knowledge (not related with time and behavior). 5) Formalize the information. For information got from step 1) to 3), we use a vocabulary to represent, while for information got from step 4), we use conditionals to represent.

Third, we construct predictive model. The predictive model represents the predictive knowledge used to make predictions. We are only interested in certain kinds of predictions, while in the target system, they are mainly accidents/attacks and other predictions, which can help to predict accidents/attacks. Therefore, the predictive model is assembled by conditionals, which can be used to get these interesting predictions. The construction of the predictive model involves following steps. 1) Determine which events/conditions (accidents/attacks and other predictions, which can help to predict accidents/attacks) we concern in the target domain, and list these events/conditions. 2) Determine the predictive knowledge related with these events/conditions. 3) Formalize the knowledge as conditionals.

Fourth, we construct behavioral model. The behavioral model is used for qualitative decision, which represents the behavioral knowledge used to choose actions. The purpose of qualitative decision is to find out "what actions should be taken?" and "which actions should be taken first?" Thus, the result of qualitative decision is a set of candidates of next actions, which are labeled as "obligatory"/"permitted" and/or priorities. Therefore, the behavioral model is assembled by conditionals, which can be used to get these results of qualitative decision. The construction of the behavioral model involves following steps. 1) List possible actions that the system can be taken to prevent accidents/attacks. 2) Decide which events/conditions cause to take these actions. 3) Formalize the information. For information got from step 1), we use a vocabulary to represent, while for information got from step 2), we use conditionals to represent.

Fifth, we evaluate the constructed anticipatory model. To evaluate anticipatory model, we could utilize some (user-defined) evaluation criteria to evaluate the model formally. Besides, in practical, we can use test set for evaluation, which is a set of empirical/historical data about accidents and attacks in the target domain.

Phase 5: Implementation
Most components of the target system are general-purpose. Thus, we do not need to rebuild them for different systems. There are some implemented components can be used in this phase, such as a general-purpose forward reasoning engine [7]. We have also implemented a general filter, a semiautomatic formula generator, a general formula chooser and a general enactor, which may deal with most cases. Our methodology considers using these prepared components. Of course, one can rebuild all of these components.

First, we implement observers or utilize some substitutes in the legacy system to observe events and conditions produced by the environment, which relate to requisite events/conditions used by anticipatory model. Second, we configure filter rules to filter nontrivial events/conditions used by anticipatory model. Third, we configure the actions mapping rules. The mapping rules map the possible anticipatory actions (based on behavioral model) are represented as logical formulas to concrete instructions that the target system can execute. Fourth, we configure translation rules according to the logic basis and anticipatory model. Fifth, we define the interesting formulas and configure the interesting terms. Interesting terms specify which terms we are interested: 1) for detection/prediction, the interesting terms are events and conditions produced by the environment about accidents and attacks, and 2) for decision-making, the interesting terms are the actions, which can be produced by the system. Interesting formula definitions specify which formulas, formalized by interesting term(s) and other symbols, we are interested. Sixth, we implement action planner and configure calculate rules. Not all target systems need quantitative calculation. Because quantitative calculation depends on the application domain, we may resort to different algorithms thus build different action planners for different target systems. Seventh, we implement the enactor and configure condition-action rules. Eighth, we integrate all components. Ninth, we test the target system. Tenth, we evaluate the target system before putting into service.

4　Case Study: Emergency Elevator Evacuation Systems

This case study is to apply our methodology to existing *emergency elevator evacuation systems* (EEES). This paper only presents important phases to apply our methodology including the result of each phase. Refer to [23] for this research's motivation, system requirements, system implementation, and system evaluation.

Phase 1: Analyze the Target Domain
The possible accidents are disasters or other bad situations, such as fire, toxic fume, biological/chemical leak, congestion, etc. The possible anticipatory actions for an EEES is to dispatch the elevator cars anticipatorily, aiming to avoid disaster beforehand and shorten the evacuation time.

Phase 2: Analyze the Legacy System
The set I includes: 1) the occupants' direct operations to the elevator system, 2) the emergency events, such as fire, bomb threats, biological or chemical threat, and 3) occupants' conditions, such as where the occupants and occupants with disabilities or injuries are. For O, the actions are mainly dispatching the elevator cars, furthermore, some elevator cars also support "shuttle mode", to avoid time needed for accelerating and decelerating smoothly [2], besides, the public address system can give warnings and instructions to occupants. The anticipatory actions are dispatching the elevator cars. The legacy system only ensures the safety of elevator cars and evacuation floors, but cannot dispatch the elevator cars anticipatorily. For I, there are many existing equipments can perceive the status of the building

Table 1 Predicate dictionary of the world model

Predicate	Meaning
$Floor(f)$	f is a floor
$Lobby(l)$	l is an elevator lobby
$Upstairs(f_1, f_2)$	f_1 is upstairs of f_2
$Locate(l, f)$	elevator lobby l locates in floor f
$Elevator(e)$	e is an elevator car
$Getatable(e, l)$	elevator e is getatable to lobby l
$Occupied(r)$	region r has alive people
$Afire(r)$	region r is afire
$Smoke(r)$	region r is full of smoke
$Person(o)$	o is a person
$Priority(o, p)$	o has priority p for rescuing
$In(o, r)$	o in the region r
$Eliminated(et)$	emergency et has been eliminated

(including elevator systems and occupants) and occupants. Emergency sensors (e.g., heat detectors, smoke detectors, flame detectors, gas sensors, etc.) can detect emergencies. The status of occupants, such as number of occupants in each floor/area, and whether there are disabled or injured people, can be monitored by using RFID [15] or cameras [11]. We can connect the filter directly to these sensors by invoking the sensors' driver program. For O, we can dispatch the elevator cars by giving instructions to elevator control system of the legacy system. When an emergency arises, the elevator control system could disable the regular dispatch mode, and receives instructions from enactor.

Phase 3: Define Requirements of the Target System
Because the purpose of emergency evacuation is to move people away from the threat or actual occurrence of a hazard immediately and rapidly, we consider the target EEES should be safe, context aware, anticipatory, and instructional/informative [23].

Phase 4: Construct Anticipatory Model
We chose *temporal deontic relevant logics* [5] as the logic basis to both predict and make decisions. Such logics satisfied the requirements of phase 4.1.

In this case, we built the vocabulary as a predicate dictionary. Table 1 shows an example. Moreover, we do not need other empirical knowledge in the world model.

In this case, the interesting predictions are emergencies. There are different empirical theories for different emergencies. For example, in fire emergency, to predict fire spread, we have following empirical theorem "the upper floor of the afire floor will catch fire with highly probable" [16, 19]: $\forall f_1 \forall f_2 (Upstairs(f_1, f_2) \wedge Afire(f_2) \wedge \neg Afire(f_1) \Rightarrow F(Afire(f_1)))$. To predict based on prediction, we have: $\forall f_1 \forall f_2 (Upstairs(f_1, f_2) \wedge F(Afire(f_2)) \wedge \neg Afire(f_1) \Rightarrow U(Afire(f_2), F(Afire(f_1))))$, which means "if a floor will catch fire, and then when this floor is really afire, its upstairs will catch fire".

Because we use temporal deontic relevant logics as the logic basis, we use "*O*" to specify an action is "obligatory" to do, and "*P*" for "permission". In order to express more precision, we also introduces priority constants (using floating number to express, smaller means higher priority): $CRITICAL > HIGH > MEDIUM > NORMAL > LOW > PLANNING$. As shown in table 2, we built the vocabulary of actions (only the last one is an explicit instruction to dispatch the elevator cars). There are three types of actions in an anticipatory EEES: reactive, anticipatory, and routine. Reactive actions deal with current crises. Anticipatory actions deal with predictive crises. Routine actions are used in a full evacuation to egress all occupants in the building, e.g., down peak. In a full evacuation scenario, the routine action could be down peak. Now we show how to construct the behavioral empirical knowledge. For example, in fire emergency, "the occupants of the fire floor are the occupants at the highest risk and should be the ones to be evacuated first by the elevators" [18], which can be expressed as empirical theorem about reactive actions: $\forall f(Afire(f) \wedge Floor(f) \wedge Occupied(f) \Rightarrow O(Pickup(f, CRITICAL, FIRE)))$. To predict crises, we have empirical theorems about anticipatory actions: $\forall f(F(Afire(f)) \wedge Floor(f) \wedge Occupied(f) \Rightarrow O(Pickup(f, HIGH, FIRE)))$ and $\forall f \forall a(U(a, F(Afire(f))) \wedge Floor(f) \wedge Occupied(f) \Rightarrow O(Pickup(f, MEDIUM, FIRE)))$. To express routine actions, e.g., down peak can be expressed as: $\forall et \forall f(FullEvacuation(et) \wedge Floor(f) \wedge Occupied(f) \Rightarrow P(Pickup(f, PriorCal(NORMAL, f), et)))$, which means when we want to apply full evacuation because of emergency type et, we can pick up each floor with priority $PriorCal(NORMAL, f)$. $PriorCal$ is a function to calculate priority, which result is $NORMAL - 0.001 \times FloorNumber(f)$, meaning higher floor has higher priority. To consider full evacuation when fire emergency, can be expressed as empirical theorem: $\forall r(Afire(r) \Rightarrow FullEvacuation(FIRE))$. Besides, because the aged and people with disabilities have higher priority in any emergency, we have: $\forall o \forall p \forall f(Person(o) \wedge Priority(o, p) \wedge Floor(f) \wedge In(o, f) \Rightarrow P(Pickup(f, p, ANY)))$. Because the elevator cars are the executor of the actions, we use $Dispatch(e, f, p, et)$ to express the action elevator e to pick up people on floor f with priority p because of emergency type et. We have: $\forall f \forall p \forall et \forall e \forall l(O(Pickup(f, p, et)) \wedge Elevator(e) \wedge Getatable(e, l) \wedge \neg IsFire(l) \wedge Locate(l, f) \Rightarrow O(Dispatch(e, f, p, et)))$ and $\forall f \forall p \forall et \forall e \forall l(P(Pickup(f, p, et)) \wedge Elevator(e) \wedge Getatable(e, l) \wedge \neg IsFire(l) \wedge Locate(l, f) \Rightarrow P(Dispatch(e, f, p, et)))$. When certain emergency was eliminated (e.g. the fire dies out), the anticipatory EEES should know the evacuation should stop, thus we have: $\forall et(Eliminated(et) \Rightarrow P(StopEvacuation(et)))$.

Phase 5: Implementation

Because we chose temporal deontic relevant logics as the logic basic, for prediction, the interesting formula (*IF*) is defined as: 1) If *A* is an interesting term, then *A* is an *IF*; 2) If *A* is an interesting term, then $\neg A$ is an *IF*; 3) If *A* or *B* is *IF*s, then $U(A, B)$ is an *IF*; 4) If *A* is an *IF*, then $\forall xA$ and $\exists xA$ are *IF*s; 5) If *A* is an *IF*, then ΦA is an *IF*, where Φ is one of $\{G, F\}$. For candidates of actions, the *IF* is defined as:

Table 2 Predicate dictionary of the behavioral model

Predicate	Meaning
$Pickup(f, p, et)$	pick up people on floor f with priority p due to emergency type et
$FullEvacuation(et)$	adopt full evacuation from the building because of emergency type et
$StopEvacuation(et)$	stop the evacuation because of emergency type et
$Dispatch(e, f, p, et)$	an action that elevator e to pick up people on floor f with priority p because of emergency type et

1) - 3) are as same as that of prediction; 4) If A is an *IF*, then ΦA is an *IF*, where Φ is one of $\{O, P\}$. For the prediction of fire, $Afire$ is an interesting term. For candidates of actions, $Pickup$, $FullEvacuation$, $StopEvacuation$, and $Dispatch$ are interesting terms.

The action planner can refine the actions by quantitative calculation, in order to plan precise elevator cars dispatch. The action planner outputs next planned actions for each elevator car. An algorithm to calculate quantitative *planned_action* refer to [23].

We have following condition-action rule in the target system: "Never stop the elevator car at an elevator lobby with an ongoing emergency."

5   Case Study: Runway Incursion Prevention Systems

This case study is to apply our methodology to existing airport/on board systems/equipments to build an *anticipatory incursion prevention systems* (ARIPS) [25]. This paper only presents important phases to apply our methodology including the result of each phase. We have submitted the this research's motivation, system requirements, system implementation, and system evaluation to some journal [25].

Phase 1: Analyze the Target Domain
The possible accidents are runway incursions and runway collisions. The possible anticipatory actions include holding a taxiing aircraft/vehicle, stopping taking off, stopping landing and going around, etc.

Phase 2: Analyze the Legacy System
The set I includes: 1) the locations of aircrafts/vehicles, and 2) the speed and acceleration of aircrafts/vehicles. These information are provided by *traffic information system* [21] and *traffic surveillance sensor system* [21]. For set O, actions are instructions, which are carried out by *human machine interface system*, which gives the alerts about runway incursions or collisions as well as the instructions for avoiding runway incursions or collisions to both pilots/drivers and air traffic controllers, and *traffic signal system*, which gives instructions to pilots/drivers by mainly using lights through different light systems deployed in the airport. The anticipatory actions are some instructions in O, such as holding a taxiing aircraft/vehicle, stopping

taking off, and stopping landing and going around. The legacy system can only give passive warnings, but cannot prevent runway incursions or collisions anticipatorily.

Phase 3: Define Requirements of the Target System
The target system should not only detect a runway incursion or predict an forthcoming collisions, but also predict a runway incursion as early as an experienced air traffic controller. Besides, the target system should give not only alerts but also explicit instructions to pilots/drivers to avoid the incursions/collisions.

Phase 4: Construct Anticipatory Model
We chose *temporal deontic relevant logics* [5] as the logic basis to both predict and make decisions.

For the world model, we built the vocabulary as a predicate dictionary (not shown in this paper). Besides, we also need some empirical knowledge, such as "if a aircraft is accelerating on a runway, then that aircraft is taking off" represented as $At(o,r) \wedge Runway(r) \wedge Aircraft(o) \wedge Accelerate(o) \Rightarrow TakeOffFrom(o,r)$.

In this case, the interesting predictions are runway incursions and runway collisions. The predictive model includes the knowledge used for generating these interesting predictions. For example, we have conditionals $C2(r_1, r_2, r_3) \wedge H(At(o,r_1)) \wedge At(o,r_2) \Rightarrow F(At(o,r_3))$ meaning "if an aircraft has crossed the hold line of a runway, then that aircraft will arrive at that runway", and $F(At(o,r)) \wedge Active(r) \Rightarrow F(RIby(o,r))$ meaning "if an aircraft will arrive at a runway and that runway is active (occupied for taking off or landing), then that aircraft will cause runway incursion on that runway".

The behavioral model specifies the actions to prevent a ongoing/furture runway incursion/collision. For example, we have conditionals $F(At(o,r)) \wedge F(RIby(o,r)) \Rightarrow O(Hold(o))$ meaning "if an aircraft will arrive at a runway and will cause a runway incursion on that runway, then the aircraft is obligatory to hold in its position", $F(LandOn(o_1,r)) \wedge F(RIby(o_2,r)) \Rightarrow O(GoAround\ (o_1))$ meaning "if an aircraft is landing on a runway and there will be a runway incursion on that runway, the aircraft should cancel landing and go around", and $TakeOffFrom(o_1,r) \wedge F(RIby(o_2,r)) \Rightarrow O(Evade\ (o_1))$ meaning "if aircraft is taking off form a runway and there will be a runway incursion on that runway, the aircraft should cancel taking off, and evade potential collision".

Phase 5: Implementation
Because we also chose temporal deontic relevant logics as the logic basic, the definition of interesting formula is as same as that in section 4. The interesting terms of predictions are $RIby$, $Collision$, $TakeOffFrom$, $LandOn$, $OccupyIntersection$, etc. The interesting terms of decision making are $GoAround$, $Evade$, $Hold$, etc.

Action planner decides an instruction for real time operation. For example, if the speed of aircraft is greater than V1, then that aircraft cannot stop taking off, because V1 is the maximum speed in the takeoff at which the pilot can take the first action (e.g., apply brakes, reduce thrust, deploy speed brakes) to stop the airplane within the accelerate-stop distance.

6 Case Study: Computing Services

This case study is to apply our methodology to existing computing services for defending against malice. This paper only presents important phases to apply our methodology to a computing service including the result of each phase. We have submitted the this research's motivation, system requirements, system implementation, and system evaluation to some conference [24].

Phase 1: Analyze the Target Domain
The possible attacks to a computing service includes DoS, probing, compromises, and viruses/worms/Trojan horses [13]. The possible anticipatory actions include blocking access from the attacker or to the target, killing connections to or from a particular host, network, and port, changing an attack's content, terminating a network connection or a user session, and blocking access form certain hosts or user accounts. Besides, the system could notify security administrators notifications about important detections/predictions/responses about malicious activities.

Phase 2: Analyze the Legacy System
The set I includes network traffic and network data. To get these traffic/data, there are two kinds of observers: inline observers and passive observers. An inline observer is deployed so that the network traffic/data transfer it is monitoring must pass through it, such as a hybrid firewall, a proxy, and a security gateway. A passive observer is deployed so that it monitors a copy of the actual network traffic, NetFlow or sFlow, host system status, and application logs and status. The actions in set O are carried out by actuators. An actuator could be a firewall (either network firewall or host-based firewall), a router, or a switch, which can block access from the attacker or to the target; An actuator could be a program that kills connections to or from a particular host, network, and port, such as Tcpkill; An actuator could be a security proxy, which can change an attack's content; An actuator could also be an application service, while some application could terminate a network connection or a user session, block access form certain hosts or user accounts.

Phase 3: Define Requirements of the Target System
The target system can identify suspicious activity, such as reconnaissance, then predict an imminent attack, besides, the system should also predict fatal attacks from trivial attacks and other suspicious activities. The target system can choose and take actions automatically to both stop an ongoing attacks and prevent an imminent attack anticipatorily. The target system should use explicable model for detecting/predicting malice and choosing actions, thus the system's behaviors are explicable. The target system can handle application level attacks. The target system can be customized for different applications and different scenarios, while interoperability is one of foundations for customization.

Phase 4: Construct Anticipatory Model
We chose *temporal deontic relevant logics* [5] as the logic basis. We present preventing web server against HTTP DoS attacks as an example. In predictive model, "$MaxConnection(ip\_addr) \Rightarrow HttpDos(ip\_addr)$" means "if a host ip_addr makes

more than max concurrent requests, it is taking a HTTP DoS attack.", and "$TooFrequent(ip\_addr) \Rightarrow HttpDos(ip\_addr)$" means "if a host ip_addr requests the same page more than a few times per second, it is taking a HTTP DoS attack." In behavioral model, $HttpDos(ip\_addr) \Rightarrow Block(ip\_addr)$ means "if host ip_addr is taking a HTTP DoS attack, block it."

Phase 5: Implementation
Because we also chose temporal deontic relevant logics as the logic basic, the definition of interesting formula is as same as that in section 4. The interesting terms of predictions are predicates of attacks, such as $HttpDos$. The interesting terms of decision making are $Block$, $Kill$, $ChangeContent$, etc.

7   Discussion

Although there are some particularities in the three case studies, we can still argue our methodology is a general one.

First, our case studies showed that our methodology can deal different kinds of reactive systems. A reactive system can have following three kinds of functions [26].

- Informative: To provide information about the subject domain, such as to answer questions and produce reports.
- Directive: To direct the entities in the subject domain, such as to control or guide a physical entity.
- Manipulative: To manipulate lexical items in the subject domain, such as to create, remove, or modify lexical items.

The legacy reactive systems in first and second case studies have directive and informative functions, while the legacy reactive system in third case study has informative and manipulative functions. Therefore, from viewpoint of system function, our case studies covers different kinds of reactive systems, thus, our methodology is general.

Second, our case studies showed that our methodology can improve both the legacy system's safety and security. *Safety* is freedom of risk [8] (accidents or losses) [14]. *Security* is prevention of or protection against access to information by unauthorized recipients or intention but unauthorized destruction or alteration of that information [8]. Safety and security are closely related and similar, while the important differences between them is that security focuses on malicious actions. Our first and second case studies improve the system's safety by avoiding disasters/accidents. Our third case study improve the system's security by preventing attacks/ malicious behaviors.

The main particularity among the three case studies is quantitative calculating of the action planer. The quantitative calculating for each case study is special, while it cannot be used in other cases. At present, we can eliminate this particularity by implementing different action planners for different target domains.

8 Concluding Remarks

It is difficult to ensure reactive systems' safety and security, because the problem is not to ensure what should happen, but to ensure what should not happen [1, 14]. Furthermore, it is more challenging to improve existing reactive systems' safety and security. We have proposed a new approach by extending the systems with anticipatory ability without reimplementing the whole of the existing system. The proposed methodology can be used to extend various reactive systems with anticipatory ability, but preserve the system's original functions.

We presented three case studies to show how to apply our methodology to legacy reactive systems. We also discussed the generality of the case studies.

We can conclude: 1) it is possible to extend an existing reactive system with anticipatory ability, 2) the anticipatory ability is useful to the legacy reactive system, thus such an extension is rewarding, and 3) our method is effective, at least in the three case studies.

To prove the generality of our proposed methodology, the best way is to apply our methodology to various target domains. Thus, our future work is to find out some other interesting areas, which require high safety or high security, then extends some existing critical reactive systems in these areas with anticipatory ability.

References

1. Anderson, R.J.: Security engineering: a guide to building dependable distributed systems, 2nd edn. Wiley (2008)
2. Bukowski, R.W.: Emergency egress strategies for buildings. In: Proc. 11th International Interflam Conference, pp. 159–168 (2007)
3. Cheng, J.: Temporal relevant logic as the logical basis of anticipatory reasoning-reacting systems. In: Proc. Computing Anticipatory Systems: CASYS - 6th International Conference, AIP Conference Proceedings, vol. 718, pp. 362–375. AIP (2004)
4. Cheng, J.: Connecting components with soft system buses: A new methodology for design, development, and maintenance of reconfigurable, ubiquitous, and persistent reactive systems. In: Proc. 19th International Conference on Advanced Information Networking and Applications, pp. 667–672. IEEE Computer Society Press (2005)
5. Cheng, J.: Temporal deontic relevant logic as the logical basis for decision making based on anticipatory reasoning. In: Proc. 2006 IEEE International Conference on Systems, Man and Cybernetics, vol. 2, pp. 1036–1041. IEEE (2006)
6. Cheng, J.: Adaptive decision making by reasoning based on relevant logics. In: Proc. Computational Intelligent: Foundations and Applications, 9th International FLINS Conference, pp. 541–546. World Scientific (2010)
7. Cheng, J., Nara, S., Goto, Y.: FreeEnCal: A forward reasoning engine with general-purpose. In: Apolloni, B., Howlett, R.J., Jain, L. (eds.) KES 2007, Part II, LNCS (LNAI), vol. 4693, pp. 444–452. Springer, Heidelberg (2007)
8. Daintith, J., Wright, E.: A dictionary of computing, 6th edn. Oxford University Press (2008)
9. Goto, Y., Kuboniwa, R., Cheng, J.: Development and maintenance environment for anticipatory reasoning-reacting systems. International Journal of Computing Anticipatory Systems 24, 61–72 (2011)

10. Harel, D., Pnueli, A.: On the development of reactive systems. Logics and Models of Concurrent Systems, 477–498 (1985)
11. Kim, J.H., Moon, B.R.: Adaptive elevator group control with cameras. IEEE Transactions on Industrial Electronics 48(2), 377–382 (2001)
12. Kitajima, N., Goto, Y., Cheng, J.: Development of a decision-maker in an anticipatory reasoning-reacting system for terminal radar control. In: Corchado, E., Wu, X., Oja, E., Herrero, Á., Baruque, B. (eds.) HAIS 2009. LNCS, vol. 5572, pp. 68–76. Springer, Heidelberg (2009)
13. Lazarevic, A., Kumar, V., Srivastava, J.: Intrusion detection: A survey. Managing Cyber Threats, 19–78 (2005)
14. Leveson, N.G.: Safeware: system safety and computers. Addison-Wesley (1995)
15. Luh, P.B., Xiong, B., Chang, S.C.: Group elevator scheduling with advance information for normal and emergency modes. IEEE Transactions on Automation Science and Engineering 5(2), 245–258 (2008)
16. Morris, B., Jackman, L.A.: An examination of fire spread in multi-storey buildings via glazed curtain wall facades. Structural Engineer 81(9), 22–26 (2003)
17. Pnueli, A., Rosner, R.: On the synthesis of a reactive module. In: Proceedings of the 16th ACM SIGPLAN-SIGACT Symposium on Principles of Programming Languages, pp. 179–190. ACM (1989)
18. Proulx, G.: Evacuation by elevators: who goes first? In: Proc. Workshop on Use of Elevators in Fires and Other Emergencies, pp. 1–13. NRC Institute for Research in Construction, National Research Council Canada (2004)
19. Quintiere, J.G.: Fire growth: an overview. Fire Technology 33(1), 7–31 (1997)
20. Rosen, R., Kineman, J.J., Rosen, J., Nadin, M.: Anticipatory systems: philosophical, mathematical, and Methodological Foundations, 2nd edn. Springer (2012)
21. Schönefeld, J., Möller, D.P.F.: Runway incursion prevention systems: A review of runway incursion avoidance and alerting system approaches. Progress in Aerospace Sciences 51, 31–49 (2012)
22. Shang, F., Nara, S., Omi, T., Goto, Y., Cheng, J.: A prototype implementation of an anticipatory reasoning-reacting system. In: Proc. Computing Anticipatory Systems: CASYS - 7th International Conference, AIP Conference Proceedings, vol. 839, pp. 401–414. AIP (2006)
23. Shi, K., Goto, Y., Zhu, Z., Cheng, J.: Anticipatory emergency elevator evacuation systems. In: Selamat, A., Nguyen, N.T., Haron, H. (eds.) ACIIDS 2013, Part I. LNCS, vol. 7802, pp. 117–126. Springer, Heidelberg (2013)
24. Shi, K., Goto, Y., Zhu, Z., Cheng, J.: An anticipatory reasoning-reacting system for defending against malice anticipatorily (submitted for publication, 2013)
25. Shi, K., Goto, Y., Zhu, Z., Cheng, J.: Anticipatory runway incursion prevention systems (submitted for publication, 2013)
26. Wieringa, R.: Design methods for reactive systems: Yourdon, statemate, and the UML. Morgan Kaufmann (2003)

Observations on Software Testing and Its Optimization

Dapeng Liu, Shaochun Xu, and Huafu Liu

Abstract. Although there are a lot of researches on software testing, most of the works are on individual testing approaches. There is no much work on the general understanding/discussion at higher level. Based on a few years' industrial experience, we reviewed some research fields of software testing, provided discussion on the issues that have emerged but not clearly represented in previous literature, and tried to establish a philosophical viewpoint of understanding software testing. By having listed some characteristics of modern software testing that have been commonly noticed though not fully studied, we pointed out that software testing is intrinsically dynamic and by nature a representative problem of multi-objective optimization. The test cases should be evolved along with the change of software. We also thought coding and testing should accommodate each other and especially coding strategy should be well chosen in favor of easy testing. A few research directions have also been pointed out.

Keywords: Software Testing, software evolution, Multi-Objective Optimization, Code to Debugging/Testing.

1 Introduction

There is no doubt that testing plays an irreplaceably critical role in software development therefor deep understanding and successful improvement on testing

Dapeng Liu
GradientX, Santa Monica, CA, USA
e-mail: dapeng@gradientx.com

Shaochun Xu
Department of Computer Science, Changsha University, Changsha, China,
Department of Computer Science, Algoma University, Sault Ste Marie, Canada
e-mail: simon.xu@algomau.ca

Huafu Liu
Department of Computer Science, Changsha University, Changsha, China
e-mail: hfliu9063@163.com

R. Lee (Ed.): *Computer and Information Science*, SCI 493, pp. 33–49.
DOI: 10.1007/978-3-319-00804-2_3   © Springer International Publishing Switzerland 2013

will help software quality. Software testing is an aggregation term which covers many different topics, such as, unit testing, smoke testing, integration testing, regression testing, etc. When we are talking about testing methods, the topic may cover many different fields [1]. Thereafter there is no single standard for all to simply compare different researches. Consequently, it is no surprise to see rarity of meditation on testing at a philosophical level. Due to recent addition of detailed researches on individual testing approaches, it seems there is a need to understand and discuss further the software testing at higher level.

There are often different opinions on same topics which may look opposite to each other while actually emphasize different aspects and complement each other in harmony. For example, while some are trying to automate testing or to create guidelines for testing efficiency, some argue that software testing asks for more prompt human intelligence; while many papers focused on how to increase the coverage of testing, some others realized that test cases can be impractically large so that tried to reduce them. Such different arguments can actually be considered simultaneously in one set of testing cases. Since designing testing strategies is a multi-objective optimization by itself, there is no single best solution.

As an augment of existing research, it seems that software testing has all of the following characteristics:

• May be different for different program frames;
• May be stately;
• Has to evolve along with development progress;
• Should be team work between development and QA;
• May be proactively integrated with development;
• May be better to be semi-automatic;
• Is a dynamic multi-objective evolutionary optimization problem.

In this work, we first try to conduct a brief survey on software testing literature, particularly those most recent one, and then with the help of our working experience in IT industry, we try to understand software testing and provide some observations at higher level that is missing in related literature. It is our hope that our observations could provide some good guidance for software development in order to better evolve and select the test cases so that automatic testing become possible. Those observations could also invoke research interests on them. It might be necessary to point out that we do not intend to conduct a complete literature survey on software testing in this paper and also we do not propose a general guideline for software testing. Moreover, we are not intended to appreciate any technical details of specific testing approaches or techniques. Instead, we focus on the ideology. As there is no panacea in software testing, we wish software development management can benefit from our discussions.

The rest of the paper is organized as follows. Section 2 discusses some recent representative publications. Section 3 presents our observations on software testing based on literature review and our working experience in industry. Some concrete or imaginary examples are given. The conclusions are given in Section 4.

2   Literature Review

In this section, we conduct a brief literature review on related fields of software testing and summarize them according to different topics.

2.1   Testing Should be Proactive, But Not Passive

Ahamed [2] expressed an interesting idea that software testing, depending on the testing method adopted, can be implemented at any time during the development process. This proposal is meaningful since a testing team or group is often separated from the development team in current IT industry. It is not rare that software developers and testers complain the gap between them.

Their idea is similar to ours [3]. However, we pushed the idea of integrating software testing into development stage at a higher level: we argued that software developers would better "*code to debugging*". Instead of occasionally considering during development process, we propose to proactively and actively take consideration of how to facilitate testing into the original system design. For example, a module reads properties from a file whose file path has been explicitly specified in source code as a constant literal. In this case, it is hard to test this module with different file contents if testing is done in parallel mode. Certainly, this challenge is conquerable since there have been a lot of effort and quite a few successful solutions [4] to dynamically change program behaviors in runtime without the need of modifying source code. However, using extra resource to test programs adds more burdens. Since dynamically changing software behavior in testing has to be done on concrete production code, tight coupling between testing code and software implementation is hurdle for software evolution.

2.2   Human Intelligence vs. Machine Automation

Itkonen et al [5] conducted a controlled experiment to compare the defect detection efficiency between exploratory testing (ET) and test case based testing (TCT). They found no significant differences in defect detection efficiency between TCT and ET. Different from traditional testing, ET emphasizes individual tester's skills during test execution and does not rely on predesigned test cases. The distributions of detected defects did not differ significantly regarding technical type, detection difficulty, or severity between TCT and ET. However, TCT produced significantly more false defect reports than ET.

2.3   Automatic Augment

Fraser and Zeller [6] proposed an evolutionary approach to improve the readability of automatic generation of test cases for programmers. Some portions of sequences of method calls are randomly selected in order to shorten call chains without altering overall coverage.

In contrast to random selection of test cases, Rubinov and Wuttke [7] presented an approach to automatically augment test suites by utilizing existing test cases to direct the generation of complex ones. In particular, they focused on automatic generation of complex integration test cases from simple unit tests. The rationale behind the approach was that since test cases contain information about object interactions, they could be used to generate new test cases.

2.4 More Is Better

It is a popular view that a high coverage on source code or its structure will potentially increase fault detection possibility. Although the size of test suites is not equal to test coverage, it is commonly considered to be closely related to testing effectiveness as well. Namin and Andrews [8] studied the influence of size and coverage on test suite effectiveness and found that both factors independently affect the effectiveness of test suites. Their experiment established accurate linear relationship between (log(size) + coverage models) and test suite effectiveness.

There are also other factors affecting the testing effectiveness. Alshahwan and Harman [9] argued that the variety of test outputs had positive relationship with test effectiveness. Thereafter they introduced a test suite adequacy criterion based on output uniqueness and formally proposed four definitions of output uniqueness with varying degrees of strictness. In a preliminary evaluation on web application testing they found that output uniqueness enhanced fault-finding effectiveness. Their study showed that the approach outperformed random augmentation in fault finding ability by an overall average of 280% in five medium sized, real world web applications. Specifically, they augmented test suites by combining traditional structural coverage criterion (branch coverage) with test cases that provide unique outputs. However, output uniqueness alone should not be used as the only criterion for selecting test cases though it is a helpful addition to traditional test generation adequacy criteria. An obvious and simple example is that two different errors that cause a web application to return a blank page would have the same output. Fortunately, checking the output of a test case is often a comparatively cheap operation which requires neither instrumentation of the code nor the building of intermediate structures, such as control flow graphs or data flow models.

Program faults with high severity often propagate to the observable output and affect user perception of an application [10]. Intuitively, the output of testing provides a valuable resource for testers to identify unexpected behaviors that are more critical from a user's point of view, especially for applications with a lot of outputs, such as web applications whose whole user interface is often a product of output. The complexity and richness of the output may most likely allow faults to be propagated to the output and therefore the testing approach is more effective.

2.5   Test Reduction

It is well known that it is hard, even impossible, to create enough number of test cases to cover complete combinations of execution branches. As a fact, most researches on automated test case generation focus on the approaches which generate test cases to increase structural code coverage. Usually, such techniques generate test cases based on information extracted from execution or source code.

While some research focuses on how to augment the size of test cases, some others try to reduce the test size. The test suite reduction techniques often aim to select, a minimal representative subset of test cases that retains the same code coverage as the test suite. However, Hao et al. [11] pointed out that test suites reduction approaches presented in previous literature lost fault detection capability. In order to avoid such potential loss, they proposed an on-demand technique to retain certain redundant test cases in the reduced test suite, which attempt to select a representative subset satisfying the same test requirements as the initial test suite. Their technique collects statistics about loss in fault-detection capability at the level of individual statements and models the problem of test suite reduction as an integer linear programming problem. They have evaluated the approach in the contexts of three scenarios.

2.6   Testing Interacting Artifacts

A multithreaded unit test creates two or more threads, each executing one or more methods on shared objects of the class under test. Such unit tests can be generated at random, but basic random tests are either slow or do not trigger concurrency bugs. Nistor et al [12] realized that testing multithreaded code was hard and expensive since a lot of tests are needed in order to achieve high effectiveness. They proposed a novel technique, named Ballerina, for automated random generation of efficient multithreaded tests that effectively trigger concurrency bugs. There were only two threads in Ballerina, each executing a single, randomly selected method, supervised by a clustering technique to reduce human effort in inspecting failures of automatically generated multithreaded tests. Evaluation of Ballerina on 14 real-world bugs from 6 popular codebases showed that Ballerina found bugs on average 2 to 10 times faster than basic random generation, and the clustering technique reduces the number of inspected failures on average 4 to 8 times smaller.

Greiler et al [13] also pointed out that testing plug-in-based systems is challenging due to complex interactions among many different plug-ins, and variations in versions and configurations. In a qualitative study, they interviewed 25 senior practitioners about how they tested plug-in applications on the Eclipse plug-in architecture. They found that while unit testing played a key role, the plug-in specific integration problems were often identified and resolved by the community.

2.7   Exceptions and Consequences

Zhang and Elbaum [14] realized that validating how code handling exceptional behavior is difficult, particularly when dealing with external resources. The reasons include two aspects: 1) A systematic exploration of the space of exceptions that may be thrown by the external resources is needed, and 2) The setup of the context to trigger specific patterns of exceptions is required. They presented an approach to address those difficulties by performing an exhaustive amplification of the exceptional behavior space associated with an external resource that is exercised by a test suite. Each amplification attempt tried to expose a program exception handling code block to new behavior by mocking normal return from an exception thrown by an external resource. The result of the experiment on available historical data demonstrates that this technique is powerful enough to detect 65% of the faults reported in the bug reports, and is precise enough in that 77% of the detected anomalies correspond to faults fixed by the developers.

Grechanik et al [15] mentioned that one important goal of performance testing is to find situations when applications unexpectedly behave for certain combinations of input values. They proposed a solution to automatically find performance problems in applications using black-box software testing. Their implementation is an adaptive and feedback-directed learning testing system that learns rules from execution traces of applications and then uses these rules to select test input data automatically for these applications to find more performance problems.

2.8   Test Data/Input

Fraser and Arcuri [16] claimed that although several promising techniques had been proposed to automate different tasks in software testing, such as test data generation for object-oriented software, reported studies only showed the feasibility of the proposed techniques, because the choice of the employed artifacts in those case studies was usually done in a non-systematic way. The chosen case studies might be biased, and thereafter they might not be a valid representative of the addressed type of software. Fraser and Arcuri evaluated search-based software testing when they were applied to test data generation for open source projects. In order to achieve sound empirical results, they randomly chose 100 Java projects from SourceForge (http://sourceforge.net/). The resulting case study was very large (8,784 public classes for a total of 291,639 byte code level branches). Their results showed that while high coverage on commonly used class types was achievable in practice environmental dependencies prohibited such high coverage.

2.9   Tests Refactoring

After Estefo [17] realized that less effort was dedicated to making unit tests modular and extensible compared to a lot of researches on software maintainability, he created a Tester (Surgeon), as a profiler for unit tests, collect information during tests execution. He also proposed a metric to measure the similarity between tests and provided a visualization to help developers restructure their unit tests.

Segall and Tzoref-Brill [18] found that Combinatorial Test Design (CTD) was an effective test planning technique that revealed faulty feature interactions in a given system. Inside the test space being modeled by a set of parameters, their respective values, and restrictions on the value combinations, a subset of the test space was then automatically constructed so that it covered all valid value combinations of every user input. When applying CTD to real-life testing problems, the result of CTD could hardly be used as it is, and manual modifications to the tests were performed. Since manually modifying the result of CTD might introduce coverage gaps, Segall and Tzoref-Brill presented a tool to support interactive modification of a combinatorial test plan, both manually and with tool assistance. For each modification, the tool can display the new coverage gaps that will be introduced and can also let users to decide what to include in the final set of tests.

3   Discussion

In this section, we will conduct a brief and enlightening discussion about some characteristics of modern software testing that have not been thoroughly concerned by academia. Here we do not frame our discussion upon any existing work; instead, we focus on these characteristics by themselves.

3.1   Different Tests Are Required by Different Programs

It has been well recognized that designing and executing test cases of different granularities and at different stages is essential to test software systems thoroughly [16]. Simple unit test cases check elementary actions to reveal failures of small parts of the system, whereas complex integration and system test cases are indispensable to comprehensively test the interactions between components and subsystems, and to verify the functionality of the whole system. Smoke test is normally used after minor system grades to see whether system can run without noticeable errors. While designing simple unit test cases is a relatively simple process that can be often automated, designing complex integration and system test cases is a difficult and expensive process and hard to automate.

```
@Test public void testIsPrime() {
    assertThat(isPrime(-2), is(false));
    assertThat(isPrime(1), is(false));
    ... ...
}
```

Fig. 1 A JUnit example

Figure 1 shows a method inside a JUnit test case used for validating if the target method returns expected results for given parameters. On the contrary, the test shown in Figure 2 focuses on the performance and assumes the validness of the tested method. While performance data measured by test cases may fluctuate according to the work load of the testing machine, it may be a good idea to run test multiple times and collect statistics. Nonetheless, to our best knowledge, there have been any such testing frameworks available in the market.

```
List<Event> events = loadEventsFromFile("history.txt");
int suc = 0, fail = 0;
TimeMarker tm = new TimeMarker ();
for (Event event : events) {
   if (rule.passEvent (event))  suc ++;
   else      fail ++;
}
System.out.println (String.format ("%d events, %d suc, %d fail, total time %s ms ",
events.size (), suc, fail, tm.getElapsedMilliSecond ()));
```

Fig. 2 Performance testing

There is not much discussion on different systems call for different testing strategies. Here are some thoughts based on our experience:

- For those distributed systems, the external call interface is clearly a weak spot for testing since incoming calls are out of control of the implementers for the internal program; therefore calling interface requires more attentions during testing.
- For real time, or latency-critical (sub-) systems, pressure test is indispensable because they are designed to work successfully under such situation. On the contrary, it is rare a necessity to test graphic user interface with continuous large amount of user operations.
- Scientific calculation should handle exceptions well and provide good performance. Due to the intrinsic complexity of such computation, algorithm design for those applications is more important than other kinds of applications and the implementation flaw is hardly tolerated.
- Even for similar applications, due to the different application environments, the importance of testing can be different. The financial report component in a banking system should be strictly tested and discrepancy of a penny in one transaction is disallowed; while the peer in online advertising industry may tolerate much serious errors because online advertisements are normally paid on 1,000 times of impression basis and business cannot really be controlled whether an ad has really been delivered due to client side configurations of web browsers. While a blinking program hiccup may make an online advertising lose thousands of advertisement deliveries, which is acceptable; however, a similar error cannot be tolerated by any bank.

As a result, it is hard to establish a consistent standard that is generally correct for whether a test has passed. Thereafter, software testing strategies and tools has to be carefully chosen to fit concrete cases. For example, in the famous Dynamo paper [19] it is mentioned *"A common approach in the industry for forming a performance oriented SLA (Service Level Agreements) is to describe it using average, median and expected variance"*, however, inside Amazon a different criterion has been used, *"SLAs are expressed and measured at the 99.9^{th} percentile of the distribution"*. Since it is expected that a sub-system may be unable to fulfill its requests occasionally, it is hard to claim there is an error in the case that a test fails. This error tolerance imposes more challenges to testing. To our best knowledge, there have been no testing tools that work with execution statistics.

3.2 Stateful

In reality sometimes it is painful to prepare target programs for testing. There are a lot of reasons. Some are listed as below:

- Data cache needs to warm up;
- One sub-system follows other sub-systems to change behavior in a distributed environment;
- Context needs to be filled in up steam;
- Some system-level real-time information is needed by the application. This implies two challenges: first, the same program should run within different environments; second, transitions between different states within operating systems may not be atomic and should be tested against.

Information shown in Figure 3 was copied from a real production log with modification to remove critical text for privacy. We can see that it took 94 seconds for the system to load data into cache before it can really start up. It is imaginable that it may sometimes be a big challenge to guide a big-data system into a desired status. Such difficulties suggest that testing should involve developers, which will be covered in Subsection 3.4.

It is well known that consistency, availability, and partition-tolerance cannot be simultaneously satisfied for a distributed system [20]; and only two of them can have first-level priority at the same time. In reality most distributed systems prefer two factors to the other one. For example, Riak [21], a distributed nosql data store system, may sacrifice consistency to some extent to promise availability and partition tolerance. In this case, a large amount of continuous writes into data storage cannot be absorbed and spread by the cluster in real time and thereafter subsequent unit tests of reading will fail because expected data are not available uniformly in the system yet. Since transitions between system states are not atomic, testing can be even harder because some system situations may be illegitimate in designer's mind.

```
00:56:04|Controller State: WAITING -> ESTABLISHED
00:56:04|Controller State: ESTABLISHED -> LOGGED_ON
00:56:04|0    QuovaReader-Loading Quova GeoPoint file.
00:56:05|466   QuovaReader-Processing Quova GeoPoint file.
00:56:46|42035  QuovaReader-Finished loading GeoPoint.
00:56:46|42038  WurflUtils-Initializeing WURFL database.
00:56:53|48944  WurflUtils-Completed Initializing WURFL database.
00:56:53|48944  WurflUtils-Building local cache of WURFL.
00:57:02|58081  WurflUtils-Finished Building local cache of WURFL data.
00:57:03|58341  enrichment.ASDB-Initializing AppSitesDB
00:57:09|64545  enrichment.ASDB-Completed Initializing AppSitesDB
00:57:09|64549  enrichment.PDB-Building Placement Lookup Tree
00:57:39|94275  enrichment.PDB-Completed Building Placement Lookup Tree
```

Fig. 3 A startup log

Even without the elusive transition between different states, testing still gets harder. For example, a well-known fact is that Java Virtual Machine (JVM) will define its host machine as running on a server or client [22]. Given the same program running on different computer configurations, the performance difference might be huge. Thereafter, in reality some companies set up testing environments that are exactly the same as their production environments. Nonetheless, since testing environment often has much less work amount than onerous production environment because testing runs predefined limited number of test cases but production environment has to handle various input coming from outside world, it is not rare that testing still misses some critical errors that can be found during implementation phase. To our best knowledge, there is no much research on the state of testing.

In some situations, a warmed up run time can be saved with the help of advanced productivity tools. JRebel [23] is a plug-in for the JVM that enables instant reloading of changes made to a Java class file without restart. Starting from version 1.4, the JVM has a hot swapping capability that allows developers to update the code on-the-fly during debugging; however hot swapping has been limited to updating method bodies only; adding methods or fields to classes is impossible. Emergence of JRebel reflects the daily needs of developers and testers for retaining software's runtime status.

3.3   Test Evolution and Decay

Software evolution is a well-observed phenomenon and software decay has become a popular research topic. Accordingly, testing techniques have to evolve according to related code changes. Since changes are mostly additional instead of substitution, test cases have to evolve to cater the new needs created by newly added code. For example, more assertions are added into existing test cases to test newly added features.

Incremental software development often implies shifting the focus of programmers. For example, for a startup online application with a few customers, testing on that application may need to focus on correctness; when it is optimized for performance later due to the growing of customer number, testing should shift the focus to concurrency. Sometimes, a rapid or continuous growth of concurrent customers urges overhaul of system framework therefore, a lot of existing code and many test cases have to be redone.

Sometimes, pure additional source code may ask for large code changes in existing test cases. One example is that a GUI program can be tested by a mouse action simulation test tool; later adding keyboard shortcut support might need the keyboard simulation functionality of the testing tool. More important, interweaving of mouse events and keyboard ones has to be tested. Previous mouse simulation may become obsolete if a user action simulator that is able to simulate both mouse and keyboard input is adopted in testing.

Sometimes, additional functionality needs large scale of code change that conflicts with previous testing. For example, a bank makes a business decision that allows its customers to overdraw from their checking account if they have saving accounts with enough money to cover the debit. This is one new (additional) service that the business provides to its customers; however, this implies reduction of source code and test cases which prevent the balance from being negative, in addition to new code that checks and synchronizes all related bank accounts to make sure there is surely enough amount saved under the customer's name.

```
@Test public void testMoney() {
    Money m = new Money ();
    m.setValue (20.00);
    try {
        m.deduct (30);
        assert (false);
    } catch (AccountViolationException e) {
        assert (true);
    }
}
```

Fig. 4 A JUnit test expecting an Exception

The above JUnit code (Figure 4) explicitly expects an exception when an operation violates the bank's account policy. Now, new additional service will break such test cases. This shows test evolution can be harder than intuitively imaged.

Similar to software evolution, some tests have to be removed due to new changes. In contrast to the hope that deleting existing code is effortless due to encapsulation of OOP, in reality it is often hard and even painful to do so. As to code development, removing existing code is often like amputating a defunct organ, in that all dependencies have to be cut and sealed or bridged. In a more common and more challenging case, an updated counterpart has to be installed in a similar location as a replacement and all previously cut execution paths have to been wired to it. Thankfully, changing testing code does not have to be so painful, especially for unit testing. Hopefully, test evolution does not bring as much challenge as code evolution because most existing tests have been decoupled decently. In other words, most different test code units do not depend on each other so that they do not form one single system to run but is just a collection of separate units, effects of changes will not undulate in large scale; Nonetheless, it is clear that test evolution is indispensable and unavoidable and thereafter it should be well studied by academic researchers.

3.4   Testing Has to Be Team Work

It is infeasible to expect an automatic test framework to be able to create or select test cases that are comparable with those designed by seasoned programmers. In industrial practice, there has never been such a useable software tool that is available with high precision. Thereafter, software developers have to take participation in designing and implementing test cases.

It is tedious for experienced programmers to devote much of their time in testing. A lot of software companies dedicate QA engineers for testing and let programmers focus on design and implementation. This might indicates black box testing is unavoidable. Since in some cases testing results have to be interpreted in specific execution background, it is impractical to expect QAs to be able to handle all challenges. Another reason for collaboration between developers and QAs has been given in Subsection 3.2 in which developers may help guide the to-be-tested system into a desired wanted state.

Software developers and QAs are expected to work closely to make testing effective and efficient. In reality, it is not rare to hear complaints from both development and testing departments about misunderstanding between them. At present, there are highly qualified software tools [24] that can help developers go through development and testing phrases. However, the adoption of such tools is backed up with development policy.

In traditional software development life cycle, such as waterfall model [25], software test only starts at a later stage. The worst part is that such arrangement prohibits early feedback of testing results into design and implementation and cut off the communication channel at an early stage between developers and QAs.

More recent software development models, such as agile scrum [26], allow feedback of testing results to be considered in software design. The philosophy behind agile is that software development is a complex procedure that is comparable with organic chemical reactions; while it has been an industrial standard routine for many years to monitor organic reactions in real time and adjust the environment accordingly, it makes sense to bring continuous feedback from coding achievement into software design and planning.

3.5   Testing Should Be Semi-Automatic

It has been well recognized that it is impossible to fully cover all aspects of a software system since there are too many possible program states. Therefore, without powerful computation ability of testing tool, it is impossible to cover enough aspects of the system.

On the contrary, as to testing, computer intelligence cannot compete with human intelligence yet. At present there is no way of discarding humane effort in designing and conducting good testing strategies, because the software systems are becoming more and more complex and evolving, and they are more often running on multiple instances of computation hardware in distributed ways. JUnit is an efficient testing framework in that human intelligence contributes test cases and the tool help running them in a semi-automatic manner. However, tools such as JUnit lacks creativeness to get heuristics from analyzing source code.

There are some researches [6][7][9] on using computer intelligence to help evolving test cases. A good testing approach should take advantage of both the fast calculation speed of computers and intrinsic intelligence of human being.

Based on our working experience, we observe that program logs, contain invaluable information about the runtime behaviors, but often are discarded. For example, logging of increased database operations either indicates increasing performance demand and thereafter calls for pressure testing or can be deemed as a signal to optimize implementation. Simple statistical report on logs can help us observe this change. There might be some reasons for this ignorance: a) log files are usually too large, b) a big software unit is often an aggregation of group work so that log files contain too many noises for any individual, c) it is not easy to adjust testing environment as wish to collect log at different severity levels since it is shared by many developers and QAs. Anyway, logged execution information has not attracted attention for directing future testing.

3.6   Code to Debugging/Testing

In the absence of fully automatic testing software tools, it is better to take testing into consideration during software development using human power [3]. Instead of occasionally considering testing during development procedure, we propose to proactively and actively take consideration of testing into the original system design. For example, a module reads properties from a file whose file path has been explicitly specified in source code as a constant literal. In this case, it is hard to test this module with different file contents if testing is done in parallel mode. A common coding paradigm to bring flexibility to the software is to put the file path into a configuration file; however, this code still needs to load information from a globally unique resource which prevents parallel and arbitrary testing. Merely adding a layer of method call makes multiple parallel testing much easier, as shown in Figure 5.

Normal code	Code to debugging/testing

{...
 	
   String fpath =
     properties.get Property ("fp");
   File f = fopen (fpath);
   read from f;
 ...
 }
	String fpath = properties.get Property ("fp");
	 loadFrom (fpath);
	 }
	
	 void loadFrom (String fp) {
	   File f = fopen (fpath);
	   //read from f;
	
	 }

Fig. 5 Psudo code showing coding to debugging/testing

Please be advised here though our example can be used to exhibit the benefit of logic encapsulation, our "code to debuggint/testing" proposal is subtly different from encapsulation. In the above example, since the file name is only used once, it is reasonable to put the literal directly in source code. (There is no reason to load data from the same multiple times.) The essence of our proposal is to offer greater flexibility to testing.

Certainly, this challenge is conquerable since there have been a lot of effort and quite a few successful solutions [4] to dynamically change program behaviors in runtime without the need of modifying source code used in production mode. Such solutions cannot be deemed as elegant as they impose tight coupling of testing coding and program implementation. This also implies involvement of QA department as early as in system design phrase. For example, a module reads properties from a file whose file path has been explicitly specified in source code as a constant literal. In this case, it is hard to test this module with different file contents if testing is done in parallel mode. Early involving into software design may give the QA department better chance to meditate future testing plans and give feedbacks to the development department for better system design. Intuitively, an elegantly designed program can be easily tested and validated because various modules are fairly decoupled. Thereafter, we should not consider involving QA in development phrase as a burden; instead, more reasonably, we should expect it to be able to foster better code quality.

Practice of code to debugging/testing has other benefits such as eliminating some troubles of testing. In the above example of loading configurations from a file whose path has been defined in source code, testing cannot be conducted in parallel mode since at any given time there is only one copy of the file content. Certainly, there have been a lot of effort and quite a few successful solutions [4] that dynamically change program behaviors in runtime without the need of modifying source code that is used in production mode. Nonetheless, being able to test programs in this way still needs extra effort.

Sometimes it is hard for QA to decide whether a sub system behaves exceptionally because in a large distributed non-critical system some program hiccups can be tolerated. However, if self-diagnosis functionality is implemented in source code directly, such as logging program's internal information, when system performance fluctuates, it would be much easier for testers to read logs than to analyze testing output. For example, a given system may consider one error in a second as acceptable but more than 10 continuous errors as an intolerable system behavior; there is no way for a human tester to run a test case more than 10 times in one second and catch the error. Maybe in the next second the system already goes back to normal status. If developers can do testing inside program itself for these abnormities, why cannot they just go one step further to self-test other errors, especially with the help of the QA crew? We consider it as a lift of the burden of software developers because debugging could be assigned back to them.

Debugging and testing are closely related. It is intuitive to imagine that a system that is easy to test can be easily debugged as well, because we can easily locate which parts run incorrectly; and verse visa, a system that is easy to debug should be easy to test, because it is ease to observe its status. Thereafter, we believe that code to debugging and code to testing are close and related slogans.

3.7   Testing is a Dynamic Multi-objective Optimization Problem

Based on above discussion, we know that testing is intrinsically dynamic in that its requirements, priorities, focuses are changing along with the software evolution. Unavoidably, some previous test cases might become obsolete while old software design decays and exiting coding assumptions are broken. Along with continuous changes, there is no way to find out one single best strategy for testing.

The design of testing should have a purpose of selecting a subset of representative test cases instead of merely augmenting or reducing the test size with consideration of resource limit, such as priority dates, computation resource, skills of available personnel. Therefore, development team may have to continuously adjust their developing and testing strategies. In abstraction, this is a representative example of dynamic multi-objective optimization with constraints. However, there has no much work to try to abstract testing and let mature optimization techniques help us decide what tests should be done and how they should be done.

Testing strategies are always made by human while multiple objectives are often simplified. For example, in fact of fast growth of customer numbers, QA department will focus mostly on the stability of the software execution. Such a fact of focusing on one target while neglecting others may have been caused by the ignorance of complex details of testing. With support of automated testing design and conduction, it should be able to fine tune the arrangement of testing resources and achieve better testing efficiency. For example, an automatic testing progress runs in background; an analyzer collected testing results and presents significant ones to QAs, who log bugs for develops through a system that provides high-level heuristics to backend testing system to adjust its testing strategies. Such a system is optimizing itself all the time. One ideal story is likely shown in the creative movie Matrix. A system should have a companying testing system that evolves along with the system itself. A system that is perfect on static theory would collapse, even brutally, someday, as the legendary Matrix version one.

4   Conclusion and Future Work

This paper has reviewed some research literatures on software testing, provided some views that have emerged but not clearly represented in previous literature, and tried to establish a philosophical viewpoint of understanding software testing. We have pointed out that software testing is intrinsically dynamic and by nature a representative problem of multi-objective optimization. We believe testing should stay as an indispensable component of software development and thereafter a deep understanding of testing would benefit software development as a whole. We also think coding and testing should accommodate each other and especially coding strategy should be well chosen in favor of easy testing. Our research could invoke research interests among those issues in software testing.

Acknowledgements. Shaochun Xu would like to acknowledge the support of Algoma University Travel and Research Fund, and the Oversea Expert Grant provided by Changsha University, China.

References

[1] Runeson, P., Andersson, C., Thelin, T., Andrews, A., Berling, T.: What do we know about defect detection methods? [software testing]. IEEE Software 23(3), 82–90 (2006)

[2] Ahamed, S.S.R.: Studying the Feasibility and Importance of Software Testing: An Analysis, arXiv:1001.4193 (January 2010)

[3] Liu, D., Xu, S., Du, W.: Case Study on Incremental Software Development. In: 2011 9th International Conference on Software Engineering Research, Management and Applications (SERA), pp. 227–234 (2011)

[4] Powermock, http://code.google.com/p/powermock/

[5] Itkonen, J., Mantyla, M.V., Lassenius, C.: Defect Detection Efficiency: Test Case Based vs. Exploratory Testing. In: First International Symposium on Empirical Software Engineering and Measurement, ESEM 2007, pp. 61–70 (2007)

[6] Fraser, G., Zeller, A.: Exploiting Common Object Usage in Test Case Generation. In: 2011 IEEE Fourth International Conference on Software Testing, Verification and Validation (ICST), pp. 80–89 (2011)

[7] Rubinov, K., Wuttke, J.: Augmenting test suites automatically. In: 2012 34th International Conference on Software Engineering (ICSE), pp. 1433–1434 (2012)

[8] Namin, A.S., Andrews, J.H.: The influence of size and coverage on test suite effectiveness. In: Proceedings of the Eighteenth International Symposium on Software Testing and Analysis, New York, NY, USA, pp. 57–68 (2009)

[9] Alshahwan, N., Harman, M.: Augmenting test suites effectiveness by increasing output diversity. In: 2012 34th International Conference on Software Engineering (ICSE), pp. 1345–1348 (2012)

[10] Dobolyi, K., Weimer, W.: Modeling consumer perceived web application fault severities for testing. In: Proceedings of the 19th International Symposium on Software Testing and Analysis, New York, NY, USA, pp. 97–106 (2010)

[11] Hao, D., Zhang, L., Wu, X., Mei, H., Rothermel, G.: On-demand test suite reduction. In: 2012 34th International Conference on Software Engineering (ICSE), pp. 738–748 (2012)

[12] Nistor, A., Luo, Q., Pradel, M., Gross, T.R., Marinov, D.: Ballerina: Automatic generation and clustering of efficient random unit tests for multi-threaded code. In: 2012 34th International Conference on Software Engineering (ICSE), pp. 727–737 (2012)

[13] Greiler, M., van Deursen, A., Storey, M.: Test confessions: A study of testing practices for plug-in systems. In: 2012 34th International Conference on Software Engineering (ICSE), pp. 244–254 (2012)

[14] Zhang, P., Elbaum, S.: Amplifying tests to validate exception handling code. In: 2012 34th International Conference on Software Engineering (ICSE), pp. 595–605 (2012)

[15] Grechanik, M., Fu, C., Xie, Q.: Automatically finding performance problems with feedback-directed learning software testing. In: 2012 34th International Conference on Software Engineering (ICSE), pp. 156–166 (2012)

[16] Fraser, G., Arcuri, A.: Sound empirical evidence in software testing. In: 2012 34th International Conference on Software Engineering (ICSE), pp. 178–188 (2012)

[17] Estefo, P.: Restructuring unit tests with TestSurgeon. In: 2012 34th International Conference on Software Engineering (ICSE), pp. 1632–1634 (2012)

[18] Segall, I., Tzoref-Brill, R.: Interactive refinement of combinatorial test plans. In: 2012 34th International Conference on Software Engineering (ICSE), pp. 1371–1374 (2012)

[19] DeCandia, G., Hastorun, D., Jampani, M., Kakulapati, G., Lakshman, A., Pilchin, A., Sivasubramanian, S., Vosshall, P., Vogels, W.: Dynamo: amazon's highly available key-value store. SIGOPS Oper. Syst. Rev. 41(6), 205–220 (2007)

[20] Gilbert, S., Lynch, N.: Brewer's conjecture and the feasibility of consis-tent, available, partition-tolerant web services. SIGACT News 33(2), 51–59 (2002)

[21] Riak, http://basho.com/products/riak-overview/

[22] The Java HotSpot Performance Engine Architecture

[23] JRebel, zeroturnaround.com, http://zeroturnaround.com/software/jrebel/ (accessed: December 15, 2012)

[24] Jira, http://www.atlassian.com/software/jira/overview

[25] Benington, H.D.: Production of Large Computer Programs. Annals of the History of Computing 5(4), 350–361 (1983)

[26] Schwaber, K.: Agile Project Management with Scrum, 1st edn. Microsoft Press (2004)

Semantic Specialization in Graph-Based Data Model

Teruhisa Hochin and Hiroki Nomiya

Abstract. This paper proposes the semantic specialization. This specialization makes it possible for a shape graph, which corresponds to a relation schema in the relational data model, to have elements and edges which are different from those of the original shape graphs, but are semantically related to them. Viewpoints (relationship lattices, respectively) are introduced as lattices of concepts for elements (edges) of shape graphs. The specialized elements (edges) are specified as common descendants of original elements (edges) in viewpoints (relationship lattices). The semantic specialization is informally and formally described. It is shown that the conventional specialization is a special case of the semantic specialization. By defining shape graphs through the semantic specialization, the semantically related elements could be handled as if these were of the original shape graphs while the elements have their own appropriate names.

Keywords: Specialization, semantic, data model, graph.

1 Introduction

In recent years, various kinds of knowledge have been represented, gathered, and used around us according to the advances of computers and computer networks. Wikipedia is an encyclopedia collaboratively created over the Internet. It gathers the knowledge of many people, and many people use it. The conceptual descriptions of web resources have been represented in the Resource Description Framework (RDF), which is a kind of semantic network. By using these descriptions, web resources could effectively be manipulated. These descriptions represent the knowledge of web resources.

The contents of multimedia data has also been represented with directed labeled graphs, which could be captured as a kind of semantic network. Petrakis *et al.* have

Teruhisa Hochin · Hiroki Nomiya
Dept. of Information Science, Kyoto Institute of Technology, Matsugasaki,
Kyoto 606-8585, Japan
e-mail: {hochin,nomiya}@kit.ac.jp

R. Lee (Ed.): *Computer and Information Science*, SCI 493, pp. 51–64.
DOI: 10.1007/978-3-319-00804-2_4 © Springer International Publishing Switzerland 2013

proposed the representation of the contents of medical images by using directed labeled graphs[1]. Uehara *et al.* have used the semantic network in order to represent the contents of a scene of a video clip[2]. Jaimes has proposed a data model representing the contents of multimedia by using four components and the relationships between them[3]. Contents of video data is represented with a kind of tree structure in XML[4].

We have also proposed a graph-based data model, the *Directed Recursive Hypergraph data Model* (DRHM), for representing the contents of multimedia data[5, 6, 7]. It incorporates the concepts of directed graphs, recursive graphs, and hypergraphs. An *instance graph* is the fundamental unit in representing an instance in the form of a graph. A *collection graph* is a graph having instance graphs as its components. A *shape graph* represents the structure of the collection graph. The schema graph has also been introduced to represent the specialization and the generalization relationships in DRHM[7]. The schema graph enables us to systematize and reuse knowledge.

Moreover, the semantic generalization has been proposed[8]. In the semantic generalization, the elements of the shape graphs, which are not the same, but are semantically related, can be generalized. The element generalized is the lowest common ancestor of the elements in a hierarchy of concepts. Viewpoints are introduced as hierarchies of concepts. Relationship trees are also introduced as hierarchies of relationships for edges. Although the generalization has been extended to use the semantics, the specialization[9, 10] is not extended yet.

This paper proposes the semantic specialization. In the semantic specialization, elements can be inherited from the original shape graphs as different elements. Although these are different from those of the original ones, these are semantically related to them. Owing to the semantic specialization, the specialized elements could have appropriate names while those elements are not independent of the elements of the original shape graphs, but are related to them. The semantically related elements are handled as if these were of the original shape graphs because the specialized elements are related to the original ones. Therefore, when the original shape graphs are the targets of retrieval, the specialized elements could also be the targets of the retrieval. It is also shown that the conventional specialization could be captured as the special case of the semantic specialization.

This paper is organized as follows: Section 2 explains DRHM. It includes the description of the semantic generalization. Section 3 explains the semantic specialization with an example. And then, the semantic specialization is formally defined. Some considerations including the merits of semantic specialization are described in Section 4. Lastly, Section 5 concludes this paper.

2 Directed Recursive Hypergraph Data Model

2.1 Fundamental Structure of DRHM

The structure of DRHM is described through examples. The formal definition is included in our previous work[6]. In DRHM, the fundamental unit in representing data or knowledge is an *instance graph*. It is a directed recursive hypergraph. It has a label composed of its identifier, its name, and its data value. It corresponds to a tuple in the relational model.

Consider the representation of the contents of the picture shown in Fig. 1(a). In this picture, a butterfly is on flowers. Figure 1(b) represents the contents of this picture in DRHM. In Fig. 1(b), an instance graph is represented with a round rectangle. For example, $g1$, $g11$, $g12$, and $g13$ are instance graphs. An edge is represented with an arrow. When an arrow has more than one element as a head or tail of the arrow, the elements are surrounded with a broken curve. For example, $g11$ is connected to $g12$ and $g13$ by the edge $e11$. An instance graph may contain instance graphs and edges. For example, $g1$ contains $g11$, $g12$, $g13$, and $e11$. These are called the constructing elements of $g1$. As a constructing element can include another element, the order of the inclusion can be considered. The instance graph $g11$ is called the first constructing elements of $g1$. If $g11$ contains another shape graph, say $g111$, $g111$ is called the second constructing element of $g1$. The ith constructing elements of an instance graph g is denoted as $V_{ce}^{i}(g)$. Please note that $V_{ce}^{i}(g)$ contains edges as well as instance graphs.

A set of the instance graphs having similar structure is captured as a *collection graph*. A *collection graph* is a graph whose components are instance graphs. It corresponds to a relation in the relational model. An example of a collection graph is shown in Fig. 2(a). A collection graph is represented with a dashed dotted line. A collection graph has a unique name in a database. The name of the collection graph shown in Fig. 2(a) is "Picture." The instance graph $g1$ is the one shown in Fig. 1(b). The instance graph $g2$ is for another picture. These instance graphs are called *representative instance graphs*.

The structure of a collection graph is represented with the graph called a *shape graph*. It corresponds to a relation schema in the relational model. The collection graph, whose structure the shape graph represents, is called its *corresponding collection graph*. Figure 2(b) shows the shape graph for the collection graph "Picture" shown in Fig. 2(a). It represents that an instance graph "picture" includes an instance

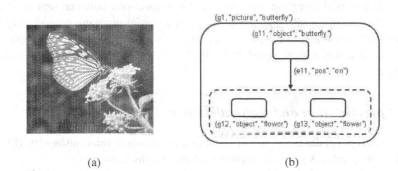

(a)   (b)

Fig. 1 A picture (a) and its instance graph (b)

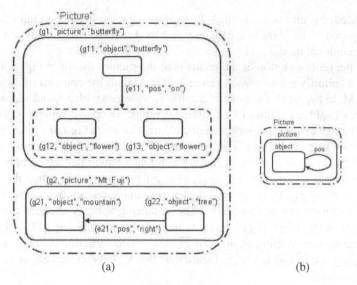

(a)   (b)

Fig. 2 A collection graph (a) and its shape graph (b)

graph "object," and an instance graph "object" is connected to "object" by an edge "pos."

The formal definition of the shape graph is presented here for the use in defining the semantic specialization. Definitions of the other graphs are included in our previous work[6].

Definition 1. A shape graph sg is a 9-tuple $(nm_{sg}, V_s, E_s, L_{v_s}, L_{e_s}, \phi_{v_s}, \phi_{e_s}, \phi_{connect_s}, \phi_{comp_s})$, where nm_{sg} is the name of the shape graph sg, V_s is a set of shape graphs included in sg, E_s is a set of shape edges, L_{v_s} is a set of labels of the shape graphs, L_{e_s} is a set of labels of the shape edges, ϕ_{v_s} is a mapping from the set of the shape graphs to the set of labels of the shape graphs ($\phi_{v_s} : V_s \to L_{v_s}$), ϕ_{e_s} is a mapping from the set of the shape edges to the set of labels of the shape edges ($\phi_{e_s} : E_s \to L_{e_s}$), $\phi_{connect_s}$ is a partial mapping representing the connections between sets of shape graphs ($\phi_{connect} : E_s \to 2^{V_s} \times 2^{V_s}$), and ϕ_{comp_s} is a partial mapping representing the inclusion relationships ($\phi_{comp_s} : V_s \cup \{sg\} \to 2^{V_s \cup E_s}$). A label of a shape graph or a shape edge is called a *shape label*. It is a triple of an identifier, a name, and a set of data types.

2.2 Specialization and Generalization

A *schema graph*[7] defines generalization or specialization relationships[9, 10] between shape graphs. A schema graph uses shape graphs as its nodes.

A specialization relationship is represented with an arrow as shown in Fig. 3(a). The shape graphs "Student" and "Teacher" are specialized from the shape graph "Person." The shape graphs "Student" and "Teacher" have the same elements of the shape graph "Person." The elements of "Student" and "Teacher" are inherited downward. In the specialization, "Student" and "Teacher" are called *specialized shape graphs*, and "Person" is called the *original one*.

A generalization relationship is represented with a broken arrow. An example of a generalization relationship is shown in Fig. 3(b). The shape graphs "Student" and "Teacher" are generalized to the shape graph "Person." The elements of the shape graph "Person" are the common elements of the shape graphs "Student" and "Teacher." The elements of "Student" and "Teacher" are inherited upward. In the generalization, "Student" and "Teacher" are called the *original shape graphs*, and "Person" is called the *generalized one*. Please note that the direction of the edge of the generalization relationship is opposite to that of the specialization relationship.

Shape graphs form a layer structure through generalization relationships and specialization ones. This structure is called the *shape graph lattice*. In this lattice, the direction of edges is the same as that of the specialization relationship. It is opposite to that of the generalization one.

It is permitted for the collection graphs corresponding to the shape graphs only at the lowest layer to have instance graphs. Let the shape graph "S" be one level higher than the shape graph "A" through a generalization relationship. For example, "Person" is one level higher than "Student" and "Teacher" shown in Fig. 3. The shape graph "S" has to be connected to the shape graph whose name is "_S" with a generalization relationship as well as "A." The shape graph "_S" has the same elements as those of "S." This shape graph "_S" is called an *exception shape graph* of "S." The collection graph corresponding to the shape graph "_S" has the instance graphs which are not included in the collection graphs whose shape graphs are at the same level in the shape graph lattice as that of the shape graph "A."

Moreover, the *data type tree* has been introduced in order to generalize data types[7]. It is a tree representing the hierarchy of data types. An example of a part of a data type tree is shown in Fig. 4. The data type *Number* is a generalized one of the data types *Int*, *Float*, *Double*, and so on. The data type of the element, whose original elements have different data types, is the one which is of the lowest common ancestor of these data types in a data type tree.

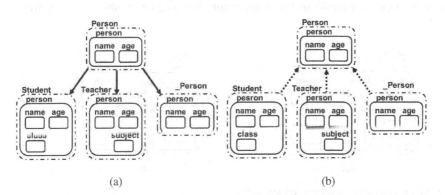

(a)   (b)

Fig. 3 Examples of specialization relationships (a) and generalization ones (b)

Fig. 4 An example of a data
type tree

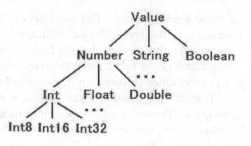

2.3 Semantic Generalization

Let us consider the two shape graphs "Soccer" and "Tennis" shown in Fig. 5. These shape graphs represent the contents of pictures of soccer and tennis games by using the notations of conceptual graphs[11]. In the semantic generalization, these shape graphs could be generalized to the shape graph "Ball_game" shown in Fig. 6, where the shape graphs "soccer_player" and "tennis_player" are generalized to the shape graph "player," and the shape graphs "soccer_ball" and "tennis_ball" are generalized to the shape graph "ball."

Semantic generalization uses[8] viewpoints and relationship trees to create shape graphs which are semantically generalized from the original shape graphs. A viewpoint is the hierarchy of concepts. Four viewpoints are shown in Fig. 7. For example, the concept "player" is the more general concept of the concepts "soccer_player" and "tennis_player." A viewpoint may have only one node. The concept "scene" is an example of this viewpoint. This viewpoint is called a *trivial viewpoint*.

A relationship tree is a hierarchy of relationships. A relationship tree is shown in Fig. 8. The relationship "relationship" is more general than the other relationships.

The shape graph "Ball_game" shown in Fig. 6 is created by applying the semantic generalization to the two shape graphs shown in Fig. 5 with the viewpoints shown in Fig. 7 and the relationship tree shown in Fig. 8.

In the semantic generalization, the correspondence between the original shape graphs and the viewpoint is specified. For example, the viewpoint, whose root is "player," is used in generalizing the shape graph "soccer_player" shown in Fig. 5(a)

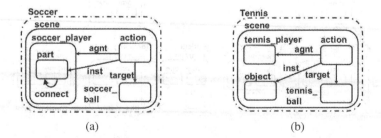

(a)   (b)

Fig. 5 Shape graphs of soccer (a) and tennis (b)

Fig. 6 Shape graph
"Ball_game" obtained by
semantic generalization

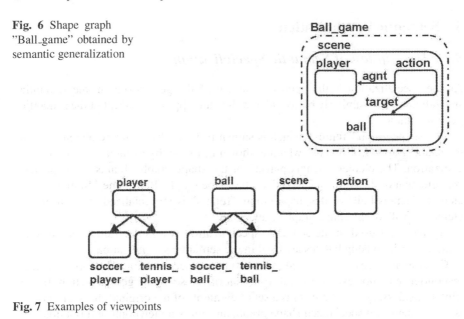

Fig. 7 Examples of viewpoints

Fig. 8 An example of a
relationship tree

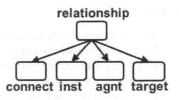

and the one "tennis_player" shown in Fig. 5(b). This correspondence is represented
with a pair < {Soccer.scene.soccer_player, Tennis.scene.tennis_player},
player >, where shape graphs are represented in the form of paths from the out-
ermost ones. Correspondences including this one are specified in the semantic generalization. By specifying the correspondence, the generalized shape graph can be
obtained as the lowest common ancestor of the original shape graphs in the viewpoint specified. The correspondences between the edges in the original shape graphs
and the relationship tree are also specified.

Specifying these correspondences is, however, considered to be cumbersome. In
our example, correspondences are not required because there is no ambiguity between shape graphs and viewpoints. The conditions of omitting the correspondences
have been examined[8]. It has also been shown that the conventional generalization
could be captured as a kind of semantic generalization[8].

3 Semantic Specialization

3.1 Description of Semantic Specialization

As the specialization is the opposite concept of the generalization, the semantic specialization can similarly be considered. It is the opposite concept of the semantic generalization.

The shape graph "Tennis," which is shown in Fig. 5(b), could be obtained from the shape graph "Ball_game," which is shown in Fig. 6, by using the semantic specialization. The element "tennis_player" of the shape graph "Tennis" is a specialized element of the element "player" of the shape graph "Ball_game." Similarly, the element "tennis_ball" of the shape graph "Tennis" is a specialized element of the element "ball" of the shape graph "Ball_game."

A viewpoint used in the semantic specialization is a lattice rather than a tree. Similarly, relationship lattices are used in the semantic specialization.

Correspondences are also used in the semantic specialization. The form of a correspondence is, however, different from that of the semantic generalization. It is a triplet $< OS, ss, vp >$, where OS is a set of elements of the original shape graphs, ss is an element of a specialized shape graph, and vp is a viewpoint or a relationship lattice. The correspondences of the semantic specialization of the example described above are as follows:

c1: $<$ {Ball_game.scene}, Tennis.scene, scene $>$
c2: $<$ {Ball_game.scene.action}, Tennis.scene.action, action $>$
c3: $<$ {Ball_game.scene.player}, Tennis.scene.tennis_player, player $>$
c4: $<$ {Ball_game.scene.ball}, Tennis.scene.tennis_ball, ball $>$
c5: $<$ {}, Tennis.scene.object, object $>$
c6: $<$ {Ball_game.scene.agnt}, Tennis.scene.agnt, agnt $>$
c7: $<$ {Ball_game.scene.target}, Tennis.scene.target, target $>$
c8: $<$ {}, Tennis.scene.inst, inst $>$

The first five correspondences are for shape graphs, while the other three ones are for shape edges. In the correspondences c1, c2, c6, and c7, the names of the elements of the specialized shape graph are the same as those of the original one. These elements are inherited from the original shape graph as in the conventional specialization. The viewpoints specified in these correspondences are trivial ones. The first items of the correspondences c5 and c8 are empty sets. These are for the specific elements of the specialized shape graph as in the conventional specialization. The names of the elements of the specialized shape graph are different from those of the original one in the correspondences c3 and c4. These elements are related to those of the original shape graph in the viewpoints specified, i.e., "player" and "ball," which are shown in Fig. 7. These are the elements inherited from the original shape graph through the semantic specialization.

Correspondences guarantee the legality of specialized shape graphs. They also represent the mapping from the original shape graphs to the specialized one.

The semantic specialization permits the multiple inheritance. That is, more than one shape graph can be original shape graphs for a specialized one. An example

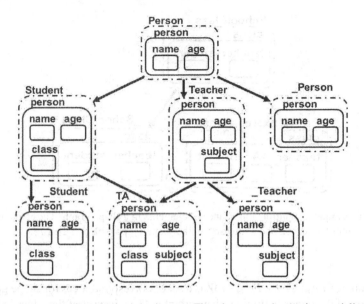

Fig. 9 An example of multiple inheritance. The shape graph TA is specialized from Student and Teacher.

of the multiple inheritance is shown in Fig. 9. The shape graph "TA" has two shape graphs "Student" and "Teacher" as the original shape graphs. In this case, for example, the correspondence $< \{\texttt{Student.person.name}, \texttt{Teacher.person.name}\}$, $\texttt{TA.person.name}$, name $>$ is required for the element "name," where the existence of the viewpoint "name" is assumed.

The multiple inheritance can be applied to the elements of a shape graph, which correspond to attributes in the relational model. An example of this kind of inheritance is shown in Fig. 10. The shape graph "ExerciseClass" is a shape graph specialized from "SchoolClass." The element "TA" of "ExerciseClass" is specialized from the elements "teacher" and "student" of "SchoolClass." The correspondence for this shape graph "TA" is $< \{\texttt{SchoolClass.class.teacher}, \texttt{SchoolClass.class.}$ $\texttt{student}\}$, $\texttt{ExerciseClass.class.TA}$, person $>$, where "person" is the viewpoint shown as in Fig. 9.

3.2 Fundamental Semantic Specialization

The semantic specialization is defined as follows.

Definition 2. In the semantic specialization, a semantically specialized shape graph sg_{term} is created. Here, $sg_{term} = SSbasic(nm_{term}, SG_{init}, T, P_{init}, P_{term}, W, Cor_p, Q_{init}, Q_{term}, R, Cor_e)$, where nm_{term} is the name of sg_{term}, SG_{init} is a set of original shape graphs, T is a data type tree, P_{init} is a set of paths of the elements of original shape graphs, P_{term} is a set of paths of the elements of a specialized shape graph, W is a set

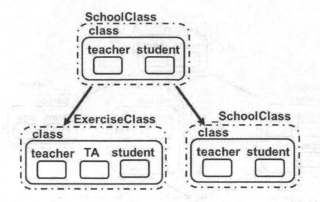

Fig. 10 An example of multiple inheritance of elements in a shape graph. The element TA is specialized from teacher and student in the shape graph SchoolClass.

of viewpoints, $Cor_p : 2^{P_{init}} \times P_{term} \to W$ is a set of correspondences for shape graphs, Q_{init} is a set of the edges in original shape graphs, Q_{term} is a set of the edges in a specialized shape graph, R is a set of relationship lattices, and $Cor_e : 2^{Q_{init}} \times Q_{term} \to R$ is a set of correspondences for edges. A specialized element (edge, respectively) must be a common descendant of the original elements (edges) in the viewpoint (relationship lattice), or the same as the original ones. The data type of a specialized element (edge) must be the data type of original elements (edges) or the one corresponding to a common descendant of the data types of original elements (edges) in T. Then, the original shape graphs are connected to sg_{term} with edges in the schema graph. A set of the exception shape graphs SG_{init}^{exc} corresponding to SG_{init} is created. Each original shape graph is connected to the exception shape graph corresponding to it, and being a member of SG_{init}^{exc} with an edge in the schema graph.

The last three steps are the same as those of the conventional specialization[7].

The function *SSbasic*, where *SS* stands for "Semantic Specialization," creates a specialized shape graph. This creation is straightforward because nothing but the shape graphs and the edges specified as the second elements of correspondences are created. Another role of this function is the confirmation of correspondences. It must be checked whether the shape graphs (the edges and the data types, respectively) specified as the original and the specialized ones in correspondences exist in a viewpoint (a relationship lattice and a data type tree), or not. Please note that the correspondences present the mappings between the elements of the original shape graphs and those of the specialized ones. This function also checks that specialized elements (edges, respectively) are common descendants of the original elements (edges) in the viewpoints (relationship lattices), or the same ones as the original ones, and that the data types of specialized elements (edges) are those of original elements (edges), or the ones corresponding to common descendants of the data types of original elements (edges) in the data type tree.

3.3 Improving Usability of Semantic Specialization

Until here, it is assumed that all of the correspondences of the elements of a specialized one are specified. An example of a correspondence of a shape graph in the conventional specialization is a triplet < {Person.person.name},Teacher.person. name, name >. This specification is cumbersome because the name of the element of the specialized shape graph is the same as that of the original one. This name could easily be obtained from the one of the original shape graph.

We introduce the function *SSext* that identifies the elements of the original shape graphs, which are not specified in the correspondences, creates the correspondences for these elements, and invokes the function *SSbasic*. As the algorithm of the function *SSext* is straightforward, the algorithm is omitted. By using *SSext*, it is sufficient to specify the correspondences truly needed. For example, specifying only the correspondences c3, c4, c5, and c8, which are described before, is enough for the semantic specialization. The remaining correspondences, i.e., c1, c2, c6, and c7 are automatically created in the function *SSext*. Please note that the correspondences for the elements "teacher" and "student" must be specified in the semantic specialization shown in Fig. 10 because these elements appear in the correspondence for the element "TA" as described before. If these correspondences are not specified in invoking the function *SSext*, the elements "teacher" and "student" will not appear in the specialized shape graph "ExerciseClass."

Invoking the function *SSext* with the correspondences for specific elements of a specialized shape graph is enough for us to use the conventional specialization. Let us consider the conventional specialization of the shape graphs "Person" and "Student" shown in Fig. 3(a). For this conventional specialization, only the correspondence < {}, Student.person.class, class > is specified in invoking the function *SSext*. This is for the specific element "class" of the specialized shape graph "Student." The other correspondences required in using the function *SSbasic*, which are described below, are automatically created by *SSext*.

< {Person.person},Student.person,person >
< {Person.person.name},Student.person.name,name >
< {Person.person.age},Student.person.age,age >

A user is required to specify nothing but the correspondence < {}, Student.person. class, class >. This specification is the same as in the conventional specialization.

4 Considerations

Specialization is an important mechanism in conceptualizing the real world[9, 10]. It can be used to define possible roles for members of a given type. Specialization has been introduced in many data and knowledge models, and programming languages. The entity-relationship (ER) model supports ISA relationships for specialization and generalization[9]. The IFO model introduces two kinds of ISA relationships: specialization and generalization relationships[10].

In the conventional specialization, the attributes of a type/class are inherited downward to the specialized type/class. For example, when the type "Dog" is defined as a specialized one of the type "Animal," all of the attributes of the type "Animal" are inherited to the type "Dog." That is, the type "Dog" has all of the attributes of the type "Animal." Moreover, the type "Dog" could have its own attributes.

Let us consider the situation that the shape graph "Tennis" shown in Fig. 5(b) is created through the conventional specification. The elements "tennis_player" and "tennis_ball" are defined as the specific elements of the specialized shape graph "Tennis." In this case, the shape graph "Tennis" has also the elements "player" and "ball" as the inherited elements. Please note that the shape graph "Tennis" has two elements on the player: "player" and "tennis_player." As the element "tennis_player" is the one that a user defined, the information on the player is inserted into this element. There are two cases whether the information is inserted into the element "player." The first case, say Case A, is that the information is inserted into it. The second case, say Case B, is that the information is not inserted into it. The same thing can be said for the elements "ball" and "tennis_ball."

When we retrieve some information on the shape graph "Ball_game," the targets of the retrieval are the shape graphs "Ball_game," "Tennis," and "_Ball_game." Some retrieval conditions on the element "player" and/or the element "ball" may be specified in the retrieval. In Case A, appropriate information is retrieved because data are inserted into the elements "player" and "ball" as well as "tennis_player" and "tennis_ball." This is, however, cumbersome because the same data must be inserted into both of the elements "player" and "tennis_player" ("ball" and "tennis_ball," respectively). As the element "tennis_player" ("tennis_ball," respectively) is also treated as the element "player" ("ball") in the semantic specialization, there is not such cumbersomeness. That is, what is needed is inserting data into the element "tennis_player" ("tennis_ball," respectively). These data are treated as if those were stored in the elements "player" ("ball").

In Case B, those conditions are meaningless for the shape graph "Tennis" because no value is stored in the elements "player" and "ball." Values are stored only in the elements "tennis_player" and "tennis_ball." This may result in false dismissal. In the semantic specialization, the element "tennis_player" ("tennis_ball," respectively) could be the target of the retrieval because it is managed with being related to the element "player" ("ball") of the original shape graph "Ball_game." This is the merit of the semantic specialization.

As described in the end of the previous section, the conventional specialization can be realized only by specifying the correspondences for the specific elements of a specialized shape graph. This specification is the same as that in the conventional specialization. The burden of this specification is equal to that of the conventional specialization. In the semantic specialization, the elements, which are semantically related to those of the original shape graphs, can be added as the inherited ones. This kind of inheritance is not supported in the conventional specialization. The semantic specialization is said to be more powerful than the conventional one. The conventional specification is considered to be a special case of the semantic one.

   Semantic specialization enables the elements of a specialized shape graph to have the names different from those of the original ones. This specialization also enables explicit specification of the inherited elements. These functions make the multiple inheritance in a shape graph possible as shown in Fig. 10. The conventional specialization does not support this kind of inheritance. This is also the benefit of the semantic specialization.

5   Conclusion

This paper proposed the semantic specialization. This specialization enables users to create the shape graphs having the elements semantically related to those of the original shape graphs. For the semantic specification, viewpoints having the lattice structure and relationship lattices are used. Correspondences between the elements of the original shape graphs and those of the specialized one are fundamentally specified. The function *SSbasic* was introduced for the fundamental semantic specialization. As specifying all of the correspondences is cumbersome, the function *SSext*, which creates the correspondences that could be generated, was also introduced. Owing to this function, users are required to specify nothing but the correspondences that could not be automatically created. The specification in the conventional specialization through the function *SSext* is the same as in the conventional specification. The conventional specification could be considered as a kind of semantic specialization.

   In the semantic generalization, the element is created as the lowest common ancestor of those of the original shape graphs in a viewpoint. On the other hand, the specialized element is created based on the specification of the semantic specialization. The difference between the action of the semantic generalization and that of the semantic specialization results in the difference of the form of the correspondence of the semantic generalization and that of the semantic specialization. Developing the framework that could treat the semantic generalization and the semantic specialization in a unified manner is in future work. Viewpoints and relationship lattices are very important in the semantic generalization and the semantic specialization. Showing the feasibility of viewpoints and relationship lattices is also in future work.

Acknowledgement. This research is partially supported by the Ministry of Education, Science, Sports and Culture, Grant-in-Aid for Scientific Research (B), 23300037, 2011-2014.

References

1. Petrakis, E.G.M., Faloutsos, C.: Similarity Searching in Medical Image Databases. IEEE Trans. on Know. and Data Eng. 9, 435–447 (1997)
2. Uehara, K., Oe, M., Maehara, K.. Knowledge Representation, Concept Acquisition and Retrieval of Video Data. In: Proc. of Int'l Symposium on Cooperative Database Systems for Advanced Applications, pp. 218–225 (1996)
3. Jaimes, A.: A Component-Based Multimedia a Data Model. In: Proc. of ACM Workshop on Multimedia for Human Communication: from Capture to Convey (MHC 2005), pp. 7–10 (2005)

4. Manjunath, B.S., Salembier, P., Sikora, T. (eds.): Introduction to MPEG-7. John Wiley & Sons, Ltd. (2002)
5. Hochin, T.: Graph-Based Data Model for the Content Representation of Multimedia Data. In: Gabrys, B., Howlett, R.J., Jain, L.C. (eds.) KES 2006, Part II. LNCS (LNAI), vol. 4252, pp. 1182–1190. Springer, Heidelberg (2006)
6. Hochin, T.: Decomposition of Graphs Representing the Contents of Multimedia Data. Journal of Communication and Computer 7(4), 43–49 (2010)
7. Ohira, Y., Hochin, T., Nomiya, H.: Introducing Specialization and Generalization to a Graph-Based Data Model. In: König, A., Dengel, A., Hinkelmann, K., Kise, K., Howlett, R.J., Jain, L.C. (eds.) KES 2011, Part IV. LNCS, vol. 6884, pp. 1–13. Springer, Heidelberg (2011)
8. Hochin, T., Nomiya, H.: Semantic Generalization in Graph-Based Data Model and Its Easy Usage. ACIS International Journal of Computer & Information Science 13(1), 8–18 (2012)
9. Silberschatz, A., Korth, H., Sudarshan, S.: Database System Concepts, 4th edn. McGraw Hill (2002)
10. Abiteboul, S., Hull, R.: IFO: A Formal Semantic Database Model. ACM Transactions on Database Systems 12(4), 525–565 (1987)
11. Sowa, J.F.: Conceptual Structures - Information Processing in Mind and Machine. Addison-Wesley (1984)

Wearable Context Recognition for Distance Learning Systems: A Case Study in Extracting User Interests

Yücel Uğurlu

Abstract. We propose a novel approach to context recognition for use in extracting user interests in e-learning systems. Under our approach, a series of screen images are captured by an imaging device worn by the user engaged in e-learning. From these images, vendor logo information is detected and used to deduce the e-learning context. The compiled history of recognized context and e-learning access can then be compared to extract low-interest topics. Experimental results show that the proposed approach is robust and able to identify the proper context 96% of the time.

Keywords: adaptive systems, distance learning, human-computer interface, wearable computing, context recognition.

1   Introduction

In recent years, distance learning systems have attracted considerable attention in higher education and e-learning in general have become an integral method for delivering knowledge. The primary advantage of e-learning is the hyper-availability of rich digital content capable of engaging students more deeply, both in and out of the classroom. With the supply of this digital content continuing to grow, e-learning systems will need to filter and adapt content to the needs and interests of the individual user [1].

Insofar as this adaptation process can be automated, by incorporating smart human-computer interaction and context recognition strategies, the e-learning experience will be improved. Indeed, the goals of next-generation e-learning systems should align well with the goals of exemplary instruction: delivering the right content, to the right person, at the proper time, in the most appropriate way—any time, any place, any path, any pace [2].

Yücel Uğurlu
The University of Tokyo, Department of Creative Informatics,
7-3-1 Hongo, Bukyo-ku, Tokyo, Japan

R. Lee (Ed.): *Computer and Information Science*, SCI 493, pp. 65–75.
DOI: 10.1007/978-3-319-00804-2_5   © Springer International Publishing Switzerland 2013

To deliver on these goals, we have focused our research on user (i.e. e-learner) context recognition based on extraction of visible information by wearable imaging devices. We believe this approach can serve most of the needs of next-gen e-learning systems, in an automated and non-intrusive way.

This paper is organized as follows: in section 2, the proposed approach is explained and a test implementation is described; in section 3, the system is applied to a sequence of test images for evaluation; in section 4, the results are assessed and future work is considered.

2 Context Recognition System

Enabling systems to act smartly and independently may the most obvious use for context awareness technology. In e-learning, context awareness can be used to decide what, when, and how learning resources are delivered to users [3]. The primary aim of our work is to discover and develop methods for recognizing a user's context from sensor data using pattern recognition.

In previous research, we developed a method for recognizing individuals' behaviors using the cameras embedded on users' PCs to analyze head postures [4]. However, understanding individuals' interest based on this approach is a challenging task, and the accuracy of the system varies significantly across changes in user race and hair color [5]. By comparison, wearable imaging devices provide much more powerful environmental coverage and are easy to integrate into many applications.

In this study, a wearable imaging device composed of a portable camera mounted on eyeglasses was used to collect images of a user's interactions with distance learning systems. The general system schema is provided in Fig. 1. As shown, context recognition provides additional information about the user's behavior and actual usage of the e-learning system, and these additional sources of information can be easily integrated into the final evaluation process.

The context recognition module is tasked with determining the user's behavior based on image processing and geometric feature matching. When using wearable imaging devices to capture images, the brightness, angle, and distance of the images change significantly as the user moves his/her head, and the captured images may also include a significant amount of noise.

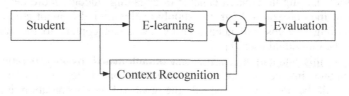

Fig. 1 An e-learning system involving context recognition

Thus, the context recognition system is decomposed into three ordered processes: image enhancement, feature detection and evaluation of the user interests.

2.1   *Image Enhancement*

By enhancing image information and restraining noise and brightness, we can significantly improve the accuracy of feature detection. The flow chart for our image enhancement process is shown in Fig. 2. The input image is used here to obtain the wavelet transform. The decomposition of images into different frequency ranges helps to isolate the frequency components introduced by intrinsic deformations or extrinsic factors. For this reason, the discrete wavelet transform (DWT) is key component of image enhancement system [6].

First, the image to be enhanced is decomposed into four frequency sub-bands using wavelet transform [7]. Second, 2D image interpolation and subtraction are applied to adjust overall image brightness. Finally, the enhanced image is obtained through the inverse wavelet transform of the sub-band images.

The resulting image has been decomposed into two parts: a low frequency part, which is a kind of average of the original signal, and a high frequency part, which is what remains after the low frequency part is subtracted from the original signal. Since the low frequency image is expected to have significant brightness variations, it should be corrected before pattern matching.

In this study, the Haar wavelet is used because it is fast and performs well at detecting sharp objects. The Haar wavelet is a sequence of rescaled "square-shaped" functions and represents the simplest possible wavelet.

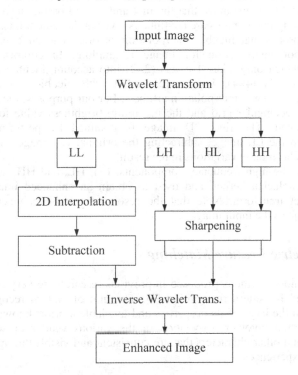

Fig. 2 Block diagram of image enhancement processes

The Haar wavelet's mother wavelet function is described as:

$$\psi(t) = \begin{cases} 1 & 0 \leq t < 1/2, \\ -1 & 1/2 \leq t < 1, \\ 0 & otherwise. \end{cases} \tag{2-1}$$

Its scaling function is described as:

$$\varnothing(t) = \begin{cases} 1 & 0 \leq t < 1, \\ 0 & otherwise. \end{cases} \tag{2-2}$$

The 2D wavelet decomposition of an image is performed by applying a 1D DWT along the rows of the image first, and then decomposing the results column-wise. This operation results in four decomposed sub-band images, referred to as low–low (LL), low–high (LH), high–low (HL), and high–high (HH). The resolution of these sub-band images is the half that of the original image.

In order to detect brightness variations in the input image, a 2D surface-fitting algorithm is employed, leveraging a bicubic spline interpolation method [8]. Bicubic splines are a method for obtaining smoothness in two-dimensional interpolation, and they guarantee that the first and second partial derivatives of the cubic interpolating polynomials are continuous, even at the data points.

In image processing, bicubic interpolation is often chosen over bilinear or nearest neighbor interpolation for image resampling. In contrast to bilinear interpolation, which only takes 4 pixels (2×2) into account, bicubic interpolation considers 16 pixels (4×4). Images resampled with bicubic interpolation are smoother and have fewer interpolation artifacts. For our purposes, the LL image is divided into a rectangular grid and the maximum brightness value for each pixel area is calculated. The final 2D image is obtained by performing bicubic interpolation on the LL image, subtracting the original LL image from that, and eliminating the brightness variations in the result.

Meanwhile, the high frequency components, LH, HL, and HH, are sharpened using a 3×3 Laplacian kernel, and used to obtain an enhanced image with the inverse wavelet transform. Note that the resolution of the enhanced image is exactly the same as the input image.

2.2 Geometric Feature Matching

Context recognition methods proposed in previous research are very similar to the techniques used in pattern recognition [9]. The aim of pattern recognition is to classify data on the basis of its properties and available a priori knowledge. In our case, classification involves association of the various vendor or school logos, unique icons, and other characters that are consistent and visible throughout most e-learning user experiences.

Pattern matching can be extremely challenging, as many factors can alter the way an object appears to a vision system. Traditional pattern matching technology relies upon a pixel-grid analysis process commonly known as normalized correlation. This method looks for statistical similarity between a gray-level model—or reference image—of an object and certain portions of the image to determine the object's position.

Though effective in certain situations, this approach may fail to find objects, or to locate them accurately, under variations in appearance common to the production or situation of those objects, such as changes in object angle, size, and shading.

To overcome these limitations, a geometric pattern matching approach is preferred. This approach involves "learning" an object's geometry, or *template*, using a set of boundary curves that are not bound to a pixel grid, and then scanning for similar shapes in the image without relying on specific grey-scale levels.

In geometric pattern matching, an object template can be found across varying lighting conditions, noise, and geometric transformations such as shifting, rotation, or scaling [10]. This makes it possible to extract the contour features of the image, by assigning each match a score that indicates how closely the template resembles the located match. This match score is given as follows:

$$r = \frac{matched\ pixels}{total\ pixels\ in\ ROI} \tag{2-3}$$

where r is the matching score and ROI is the region of interest. The output indicator of the found number of matches would indicate the number of exact matches of the template found in the database.

The result is a revolutionary improvement in the ability to accurately find objects despite changes in angle, size, and shading. The correlation coefficient has the value $r = 1$ if the two images are absolutely identical, $r = 0$ if they are completely uncorrelated, and $r = -1$ if they are antithetically uncorrelated, such as when one image is the negative of the other. In this study, the scores greater than 0.5 ($r \geq 0.5$) considered as significant correlation and the images are treated as successful matches.

2.3 Evaluation of User Interests

Integration of e-learning data with context recognition results is essential to evaluating user interests and adapting the e-learning system to those interests. Only utilizing the user access information and time is not enough to adaptive systems. The relationship between e-learning content and user reaction can be measured and can be used for the further analysis and evaluation. For this purpose, e-learning access and context recognition results are converted to numerical data such as 1 and 0 which represents valid e-learning usage or context recognition is case of 1, and return to 0 in case of failure.

E-learning access history can be represented numerically using

$$b_t = \begin{cases} 1 & access\,confirmed, \\ 0 & otherwise. \end{cases} \qquad (2\text{--}4)$$

where t is time, and b_t returns to 1 if user access is confirmed.

The context recognition data of a user is calculated using

$$d_t = \begin{cases} 1 & r_1 \geq 0.5\,or\,r_2 \geq 0.5, \\ 0 & otherwise. \end{cases} \qquad (2\text{--}5)$$

If $b_t = d_t$, e-learning access is correlated with the context recognition data for the specified access time, indicating that e-learning is taking place at that moment.

Lower interest topics can be extracted using the following equation.

$$m_t = 1 - (b_t - d_t) \qquad (2\text{--}6)$$

When $m_t = 0$, there is a good match between e-learning usage and actual context awareness, but when $m_t = 1$, the user likely has low interest in the presented content, or has lost motivation for other reasons.

Note that $m_t = 2$ is an invalid value, because it can only occur when $b_t = 0$ and $d_t = 1$, i.e. when no e-learning history has been recorded in the system.

$$n_t = mode(m_{t-2}, m_{t-1}, m_t, m_{t+1}, m_{t+2}) \qquad (2\text{--}7)$$

Since our pattern recognition method is not perfect, false detections can occur in d_t data points, yielding misleading results. To eliminate these false detections, sequential image information is used, per Equation (2–7). The "*mode*" is the value that occurs most often and in the majority of the neighboring values. Thus, false detections in up to 2 sequential images are corrected using their neighboring data points.

In case of $n_t = 0$, e-learning access history and context recognition results do not match each other and this implies a loss of user interest. In other words, even though user has an access to the e-learning system, context recognition is failed by detecting the logo information in the screen images.

3   Experimental Results

To measure the performance of the proposed system, we tested it on users of National Instruments' e-learning portal. This portal consists of more than 200 topics and 30 h of learning time, all dedicated to teaching graphical programming in LabVIEW to engineering students. The e-learning content includes various one-topic videos, each from 5 to 10 minutes in length, readings, and short quizzes. The main topics covered are:

- LabVIEW programming I
- LabVIEW programming II
- Data acquisition and analysis
- Field-programmable-gate-arrays (FPGA)

- Real time programming
- Image acquisition and processing

In practice, a unique user ID and password are provided to access all e-learning functions. Since users are logged into the system using their user IDs, it is easy to acquire their access history, as formalized in Equation (2–4).

After login, we captured a series of e-learning screen images from the imaging device worn by the user. Each image was saved to the system with a timestamp for use in further evaluations. 76 images in total were captured, at a resolution of 1280×1024 pixels.

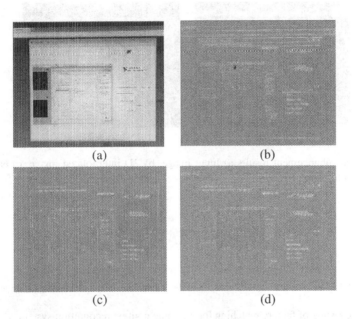

(a)   (b)

(c)   (d)

Fig. 3 Haar Wavelet transform and its sub-band images for a typical e-learning screen: a) Low-low, b) High-low, c) Low-high, d) High-high

The image enhancement and geometric feature matching processes were applied to all 76 screen images, among which there were 64 images of displayed e-learning content and 12 images of the computer display without e-learning content.

All input images used in this paper are gray scale images. For image processing, NI LabVIEW and IMAQ Vision Development software were used to analyze the screen images [11]. In LabVIEW, the *Multiresolution Analysis 2D.vi* function was used to acquire the wavelet transform images. Fig. 3 shows the wavelet transform of an image for LL, HL, LH and HH sub-bands.

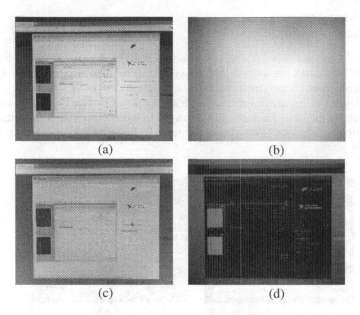

(a)   (b)

(c)   (d)

Fig. 4 Image enhancement results a) Input image, b) 2D interpolated image, c) enhanced image, d) inversed image

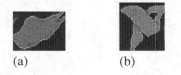

(a)   (b)

Fig. 5 Vendor logos and their contour detections a) Aqtair logo, b) NI logo

Table 1 The results of feature matching for wearable context recognition system

Number of images	Actual	Detected	Accuracy
E-learning related images	64	62	97%
Other context images	12	11	92%
Total images	76	73	96%

Since the images were divided into different frequency bands, the LL images included the brightness variations and needed to be adjusted prior to geometric pattern matching.

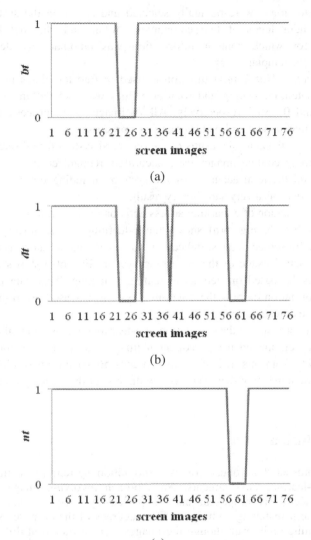

(a)

(b)

(c)

Fig. 6 Practical usage of context recognition for distance learning systems (a) e-learning access history, (b) context recognition data, (c) detection of low interest topics from screen images

Fig. 4 shows the image enhancement results using the *2D Interpolate.vi* function. For this, the original images were divided into a 5×4 grid and the maximum brightness level for each rectangular area was found. The *2D interpolate.vi* function accepts tabulated X, Y, and Z values and provides interpolated values z_i that correspond to each x_i, y_i location.

Next, vendor logos were manually selected and saved on the system. Fig. 5 shows the vendor logos of two companies, Aqtair and National Instruments Corporation, for which contour information was automatically detected and highlighted in the template images.

The *Geometric Matching.vi* function is used to find templates in LabVIEW. For pattern matching, scaling and rotation values were set within the ranges of 50%–150% and 0°–360° respectively. All 76 images were processed using the same image enhancement and pattern matching processes.

Table 1 shows the analysis results of the proposed system for all screen images: 97% of e-learning context images were accurately recognized using vendor logo information, and the total accuracy of the system, including other contexts, was 96%. We consider this a very satisfactory result.

A further evaluation of e-learning access and context recognition is provided by the graphs in Fig. 6. Fig. 6(a) shows the e-learning access history of a user throughout the 76 screen images, indicating that access history alone is not enough to understand actual usage of the e-learning system. Fig. 6(b) shows the context recognition results based on geometric feature matching. This data provides us additional information on how the e-learning system was used, even though there are several false detection points.

Finally, Fig. 6(c) shows the correlation of e-learning access and real usage, with false detections eliminated using sequential image information, as formalized in Equation (2–7). From this, it is clear that images 56–61 are uncorrelated with e-learning access, suggesting that user has low interest in the topics presented during that interval.

4 Conclusion

This paper presented a robust context recognition system for extracting user interests in e-learning environments. An approach involving image capture by wearable devices was proposed, and supported by image enhancement and geometric pattern matching techniques. The accuracy of the proposed system was found to be quite high, even though the images were acquired at different times, under different lighting, orientation, and distance conditions.

In addition, e-learning access history and context recognition data were compared to extract low interest topics from the image sequences. This information appears useful to improving e-learning contents and building more adaptive learning systems.

Future studies will focus on understanding the main characteristics of low interest topics as well as the relationship to personal needs. For this purpose, a larger dataset comprising multiple users should be collected and analyzed to identify similarities and differences between users and their behavior.

Acknowledgment. This work was supported by National Instruments Japan Corporation.

References

1. Clark, R.C., Mayer, R.E.: e-Learning and the Science of Instruction. Pfeiffer (2007)
2. Shute, V., Towle, B.: Adaptive e-learning. Educational Psychologist 38(2), 105–112 (2003)
3. Blum, M.L.: Real-time context recognition. MIT Media Lab (2005)
4. Ugurlu, Y.: Human interactive e-learning systems using head posture images. Proc. World Academy of Science, Engineering and Technology 65, 1107–1110 (2012)
5. Ugurlu, Y.: Head posture detection using skin and hair information. In: Proc. Int. Conference on Pattern Recognition, pp. 1–4 (2012)
6. Demirel, H., Anbarjafari, G.: Image resolution enhancement by using discrete and stationary wavelet decomposition. IEEE Trans. on Image Processing 20(4), 1458–1460 (2011)
7. Vetterli, M., Kovacevic, J.: Wavelets and Subband Coding. Prentice Hall (1995)
8. Kouibia, A., Pasadas, M.: Approximation of surfaces by fairness bicubic splines. Advances in Computational Mathematics 20, 87–103 (2004)
9. Duda, R., Hart, P., Stork, D.: Pattern Classification. Wiley, New York (2001)
10. Sharma, G., Sood, S., Gaba, G.S.: Image recognition systems using geometric matching and contour detection. International Journal of Computer Applications 51(17), 49–53 (2012)
11. Klinger, T.: Image Processing with LabVIEW and IMAQ Vision. Prentice Hall PTC (2003)

An IDS Visualization System for Anomalous Warning Events

Satoshi Kimura and Hiroyuki Inaba

Abstract. Recently, illegal access to the network is increasing. It has been a serious problem. To deal with this problem, necessity of Intrusion Detection System(IDS) is increasing. IDS is the notifying system of network manager to inspect symptoms of the illegal access. IDS enables us to early detect threatening attack to the computers and to deal with its attacks. However there is a problem of IDS. It is tremendous warning logs especially for large scale network. Analyzing these logs apply a large amount of load to a network manager. To overcome this problem, there exist several methods for analyzing logs based on past tendency and some visualization methods for the logs. In this paper, we propose a novel visualization system of IDS considering order relation of IP addresses that emphasize the anomalous warning events based on past tendency.

1 Introduction

Recently, as Internet is developing explosively, illegal access to the network is increasing. It is a serious problem. To deal with this problem, the necessity of Intrusion Detection System(IDS) is increasing. IDS is the notifying system of network manager to inspect symptoms of the illegal access. IDS also enables us to early detect threatening attack to the computers and to deal with its attacks. However, there exists a problem of IDS. While IDS generally outputs the text warning logs, the amount of these logs are tremendous especially for large scale network. Analyzing these logs apply a large amount of load to a network manager. To overcome this problem, there exist several methods for analyzing logs based on past tendency[1, 2, 3] and some visualization methods for the logs[4, 5, 6, 7].

In this paper, we propose a novel visualization system of IDS considering order relation of IP addresses that emphasize the anomalous warning events based on past

Satoshi Kimura · Hiroyuki Inaba
Computer Science, Kyoto Institute of Technology, Kyoto, Japan
e-mail: kimura08@sec.is.kit.ac.jp, inaba@kit.ac.jp

R. Lee (Ed.): *Computer and Information Science*, SCI 493, pp. 77–90.
DOI: 10.1007/978-3-319-00804-2_6   © Springer International Publishing Switzerland 2013

tendency. Visualization on emphasizing anomalous warning events enable network manager to reduce the analyzing load than conventional methods.

We utilize Snort IDS[8] as IDS. Snort is free and open source software. That is generally classified signature match IDS.

2 Related Work

Research on analyzing network packets and its visualization have been studied intensely for many years. In this chapter, we introduce several works.

2.1 Works on Log Analysis

We introduce several works on log analysis of IDS. There exists a study on determining theoretical statistical distribution from observation of real IDS logs[2]. They pay attention to the frequency of detections of events in an unit time, and they newly introduce three parameters. Those parameters are frequency of detections of events in an unit time, arrival time and event vast length of each events. As a result, the frequency of detections can be represented by poisson distribution, and arrival time and event vast length can be represented by exponential distribution. Therefore, they confirm that each event randomly detected at a glance can be represented by theoretical statistical distribution. They also say that the detected events out of these theoretical statistical distributions are regarded as anomalous events.

In [3], a log analysis method of paying attention to variation of detected events and its frequency in IDS logs are proposed. They plan to detect anomalous events from the number of detected events. To do their aim, they use frequency analysis method which pays attention to the variation of the number of detections, and a ratio analysis method which uses the average and a standard deviation for each event. As a result, they could specify anomalous warning events of tremendously increasing events even though the number of detections is small.

2.2 Work on Visualization

In the following, we will explain Hashing Alert Matrix[4] as log visualization system HeiankyoView[5] and IDS RainStorm[6] as log visualization system. Hashing Alert Matrix[4] is the visualization method of IDS data which use a hash function. In the method, IP address of each event is not plotted as the raw value, but it is plotted as the hash value. By using the hash value, it can use a limited drawing area efficiently. However, by using the hash value, the hierarchical structure of IP address is disappeared.

HeiankyoView[5] is the visualization method of network intrusion detection data which pay attention to hierarchical structure of IP address. Since it can arrange all detected data on two]dimensional plain smartly, we can grasp the situation of detections at a glance.

IDS RainStorm[6] is the visualization method of IDS data which is specialized in displaying class B of IPv4. It also has the property that we can grasp relation of source and destination IP address along time axis.

3   Visualization System Considering Order Relation of IP Addresses That Emphasize the Anomalous Warning Events

In this chapter, we propose a new method of visualization system considering order relation of IP addresses that emphasize the anomalous warning events by using the improved version of our proposed method for log analysis[1], and UnEqual Scaling Alert Matrix[7].

3.1   Log Analysis Method Using the Coefficient of Variance

We have studied the log analysis method using the coefficient of variance without relying on a network manager subjectivity[1]. First, we compare the number of detections N_1 during the most recent term T_1 with the number of detections N_2 during the its previous term T_2 for each event. The term T_1, T_2 are called intensive terms. If a certain ratio F is greater than a certain threshold T_h, then a warning message is outputted. We have defined a ratio of the number of detection F as following formula.

$$F = \frac{(N_1 - N_2)}{N_2} \tag{1}$$

The length of the intensive term is dynamically decided by the past three months log data. The image of the method is shown in Fig.1.

To decide the length of the intensive term, we first calculate a standard deviation for each event. Since an average number of detection is considerably different for each warning event, a standard deviation itself is not suitable for deciding the intensive term. Therefore, we use a coefficient of variance which is a value of standard deviation divided by average number of detection. It is suited for a characteristic measure for deciding the intensive term which is not dependent on the number of each warning event. We decide coefficient of variance $C_v(T)$ for intensive term T as following formula.

$$C_v(T) = \frac{S_d(T)}{\overline{X}(T)} \tag{2}$$

In the formula, $S_d(T)$ is a standard deviation, and $\overline{X}(T)$ is an average number of detections. Generally, it is expected that a variation of coefficient of variance decrease with increasing intensive term. We define a rate of change $R_d(T)$ associated with intensive term T as following formula.

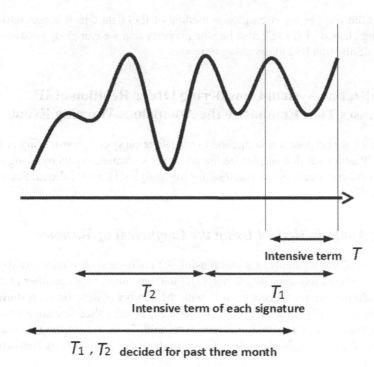

Fig. 1 Image of deciding intensive term

$$R_d(T) = \left| \frac{C_v(T+1) - C_v(T)}{C_v(T)} \right| \tag{3}$$

The intensive term T is decided such that the rates $R_d(T) \leq R_{th}$. Where R_{th} is a threshold value, 0.01 is used in [1].

Since a value of coefficient of variance indicates relative dispersion of each warning event, it is expected that a large threshold value T_h is required for a large coefficient of variance. Therefore, we define the threshold T_h as following formula.

$$T_h = \delta \times C_v(T) \tag{4}$$

The parameter δ is constant value which does not depend on a kind of warning events. As a result of experience, the value of coefficient of variance is near 1.0. By increasing δ every 0.1 used in Eq.(4), we could confirm that the number of the detections over threshold value steadily decreased.

A best of features in [1] is that the threshold of each signature is dynamically determined by coefficient of variance without a network manager's subjectivity. However, it should be noted that the method make an assumption that the number of detections for deciding intensive term T is stable.

By the way, if the value of N_1 is lower than N_2 in Eq.(1), we may also define the situation as anomalous one. But we consider this case as non-anomalous warning event in this study.

3.2   Methods of Improved UnEqual Scaling Alert Matrix

In order to grasp the information of source and destination IP address of each warning event at single display, let us prepare the two-dimensional plain having a horizontal axis as source IP address and a vertical axis as destination IP address.

However, this method has the fatal problem. We can not grasp the situation of warning events correctly because many different points are plotted on the same point of a monitor having general resolution about 1024 pixel × 1024 pixel. To overcome this problem, the authors have proposed UnEqual Scaling Alert Matrix[7]. In the method, all 32bit IP address is unequally plotted according to the ratio of the number of source and destination IP block. An outline of this method is as following.

1. The data of logs for certain fixed interval are sorted in ascending order on source or destination IP address.
2. The sorted data are classified by the first octet of IP address. After that they are hierarchically classified by the second octet and the third octet in the same way(Fig.2).
3. A plotted area of display is divided based on the ratio of each classified octet. Therefore, an IP block which is observed more frequently have a wider plotted area(Fig.3).

It is noted that the order of IP address is preserved in this method. Therefore, a network manager can easily grasp the situations of several attacks such as network scan which is observed under some IP blocks.

In order to emphasize anomalous warning events, they are plotted by more remarkable points than other normal events. The anomalous warning events are decided by the log analysis method using coefficient of variance denoted in section 3.1. In our implementation, the anomalous warning events are plotted by 5 × 5 dots larger red points and the normal warning events are plotted by 3 × 3 dots small points. Furthermore, priority 1's warning events are plotted in black color, and priority 2's warning events are plotted in green color. Also priority 3's are plotted in blue color. Where, priority 1-3 are attack risk level which are set for each Snort's signature. By this color classification, it is expected that a network manager can pay attention to each warning event appropriately. In order to grasp temporal change of detected warning events, we also implement animation mode which can continuously draw the situation every one hour.

Figure 4 shows a display image of the visualization system considering order relation of IP addresses that emphasize the anomalous warning events (A part of the emphasized event is enlarged). An area of drawing region is 800 × 800 dots, and a number of detections which are plotted in Fig.4 is 8982. To easily grasp hierarchical structure of IP address, only the scale of the first octet of source and destination IP address is drawn. When a network manager click to the plotting point that he pay

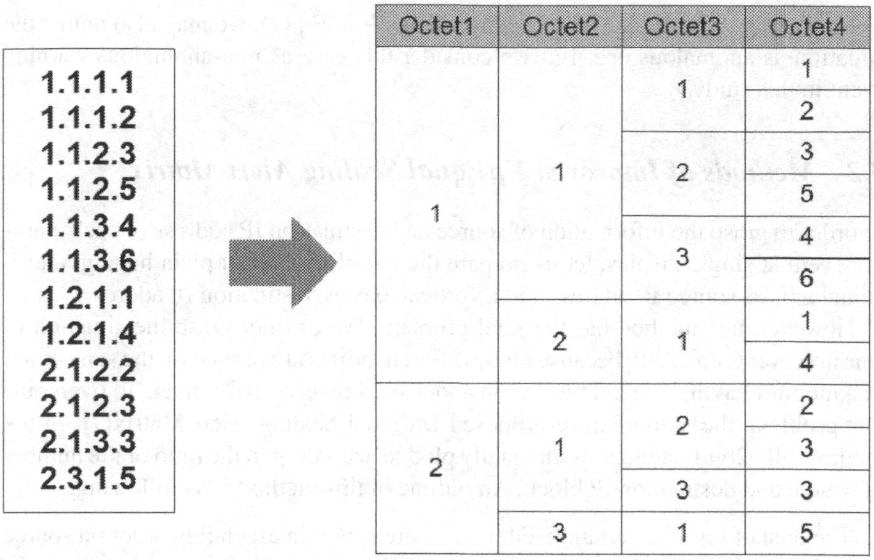

	Octet1	Octet2	Octet3	Octet4
1.1.1.1			1	1
1.1.1.2				2
1.1.2.3		1	2	3
1.1.2.5	1			5
1.1.3.4			3	4
1.1.3.6				6
1.2.1.1		2	1	1
1.2.1.4				4
2.1.2.2			2	2
2.1.2.3	2	1		3
2.1.3.3			3	3
2.3.1.5		3	1	5

Fig. 2 Classification of observed IP address

attention to, the detailed information of the warning events is displayed on a terminal. When two or more plotting points overlap each other, all detailed information for each warning event are displayed on a terminal (Fig.5).

4   Evaluation of Proposal Methods

First, we show experiment condition for evaluating the proposal methods. Secondly, we show the experimental result about the number of anomalous warning events, and show the display image of the visualization system which emphasize anomalous warning events.

4.1   Experiment Condition

The proposed visualization system is implemented by Java programming language. The detected data for evaluation is obtained from our campus network(IPv4 B-class network). The term of getting data is from 2012/5/1 through 2012/5/7. These seven days are the real observed term to evaluate the number of anomalous warning events. The term for deciding intensive term T is three months for 2012/2/1 through 2012/4/30. These three months are learning term to decide intensive term T for each signatures. Intensive term T has discrete value.

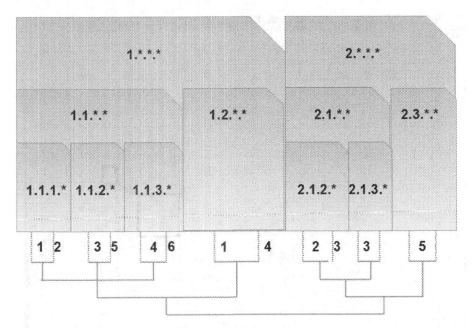

Fig. 3 Divide based on the ratio of each classified octet

4.2 Evaluation of the Number of Anomalous Warning Events

Fig.6 is the relationship between the number of anomalous warning events and the parameter δ defined in Eq.(4).

In Fig.6, we can see that the number of detections decreases with the parameter δ increasing. On deciding anomalous warning events, we previously exclude the event signatures detected less than one per hour on the average. Because it is considered that the rare event signatures are meaningless in statistics. If the rare events are detected, we can consider the events anomalous. After excluding the rare events, we call the remaining signatures target signatures. There exist 25 target signatures in our experiment.

We show some examples of target signatures in Table.1. It is considered that these target signatures are suitable as comparable subject. Because we pay attention to the following points.

- Coefficient of variance $C_v(T)$
- A ratio of anomalous warning events decreasing with a value of parameter δ
- Average detections
- Intensive term T

Number of average detections in Table.1 means the number of average detections per hour for the target signatures. $C_v(T)$ is the value of coefficient of variance for the intensive term T decided in Eq.(3).

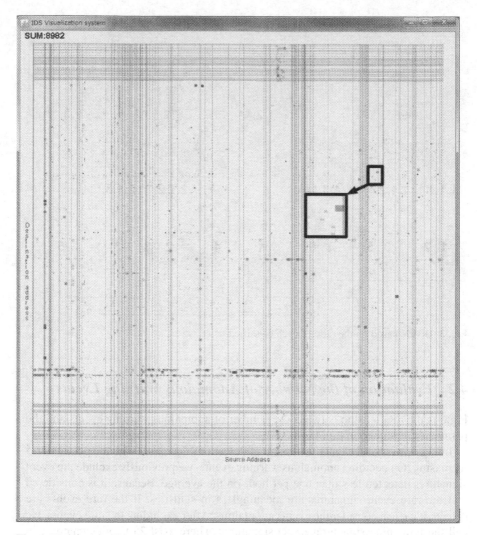

Fig. 4 UnEqual Scaling Alert Matrix which emphasize anomalous warning events and a part of the enlarged view

The event name correspond to signature number in Table.1 are described in Table.2. The decreasing rate of the number of detections for signature number 7, 8, 11, 22 in Table.1 are shown in Fig.7.

In Fig.7, the number of anomalous warning event having small value of coefficient of variance such as signature number 7, 8 tremendously decreases with parameter δ increasing. In contrast, the number of anomalous warning event having large value of coefficient of variance such as signature number 11, 22 hardly decreases with parameter δ increasing. It is considered that the signatures having small value

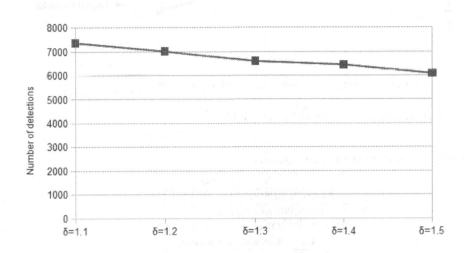

```
[2012-05-02 15:13:35.0] {WEB-CGI count.cgi access} {S_IP=         } {D_IP=         } (P:2)
[2012-05-02 12:53:16.0] {SNMP request tcp} {S_IP=         } {D_IP=     } (P:2)
[2012-05-02 12:53:16.0] {SNMP request tcp} {S_IP=         } {D_IP=     } (P:2)
[2012-05-02 20:31:12.0] {WEB-MISC robots.txt access} {S_IP=     } {D_IP=     } (P:2)
[2012-05-01 06:28:07.0] {WEB-MISC robots.txt access} {S_IP=     } {D_IP=     } (P:2)
[2012-05-02 20:31:12.0] {WEB-MISC robots.txt access} {S_IP=     } {D_IP=     } (P:2)
[2012-05-01 06:28:07.0] {WEB-MISC robots.txt access} {S_IP=     } {D_IP=     } (P:2)
[2012-05-02 20:31:12.0] {WEB-MISC robots.txt access} {S_IP=     } {D_IP=     } (P:2)
```

Fig. 5 Detailed information on terminal

Fig. 6 Number of anomalous warning events

Table 1 Detailed information for the major target signatures

Signature No.	Number of average detections	Intensive term T	$C_v(T)$	Number of detections				
				$\delta = 1.1$	$\delta = 1.2$	$\delta = 1.3$	$\delta = 1.4$	$\delta = 1.5$
7	3.8	10	0.182	68	56	42	29	23
8	32.4	11	0.161	577	542	385	358	307
11	5.7	7	1.558	639	631	611	601	601
22	2.2	15	1.279	449	447	424	409	408

of coefficient of variance are detected stably, and the signatures having large value of coefficient of variance are detected unstably for the past three months.

We show the time variance of the number of detections for signature number 8 and 11 every 1 hour in Fig.8 and 10. We also show a ratio F defined by Eq.(1) for signature number 8 and 11 in Fig.9 and 11.

A broken line in Fig.9, Fig.11 indicates the threshold value T_h defined by Eq.(4). The parameter δ in Eq.(4) is set to 1.1.

In Fig.8, we can see that the warning event(No.8) is continuously detected(Note that No.8's $C_v(T)$ is small). On the other hand, the warning event(No.11) is not

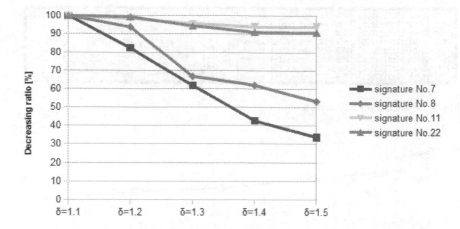

Fig. 7 Decreasing rate of signature No.7, 8, 11, 22

Table 2 Name of the major target signatures

Name of signature
7: ATTACK-RESPONSES 403 Forbidden
8: WEB-MISC robots.txt access
11: DNS SPOOF query response with TTL of 1 min. and no authority
22: SNMP request tcp

continuously detected(Note that No.11's $C_v(T)$ is large). We confirm that the other target signatures tend to the same property. These observations show the trueness of our previous consideration.

In Fig.9, a ratio F and the threshold T_h for signature 8 is shown. Each area where a solid line is over a broken line indicates the anomalous situation. Since the time variance of the number of signature number 8 in Fig.8 is stable, this diagram in Fig.9 slowly changes. So a ratio of the number of anomalous warning event becomes larger with the parameter δ increasing. On the other hand, since the time variance of the number of signature number 11 in Fig.10 is unstable, this diagram in Fig.11 suddenly changes. So a ratio of the number of anomalous warning event becomes smaller with the parameter δ increasing.

Finally, we show the original UnEqual Scaling Alert Matrix visualization system in Fig.12 and the improved version of the method in Fig.13.

In Fig.12, anomalous warning events are not emphasized and all warning events are plotted with equal shape. On the other hand, in Fig.13, anomalous warning events are emphasized.

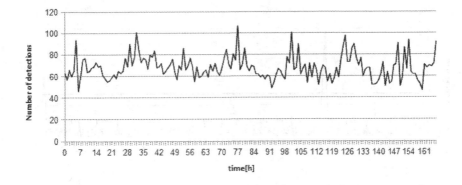

Fig. 8 Number of detections for signature number 8

Fig. 9 Increasing ratio F of the number of detections for signature number 8

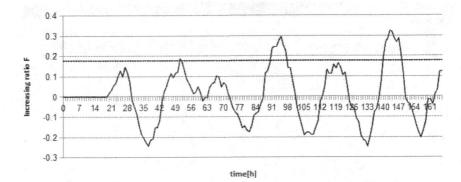

Fig. 10 Number of detections for signature number 11

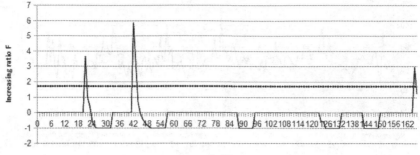

Fig. 11 Increasing ratio F of the number of detections for signature number 11

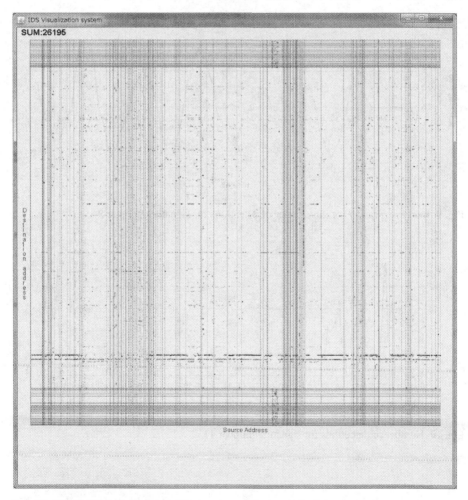

Fig. 12 UnEqual Scaling visualization system without emphasizing anomalous events

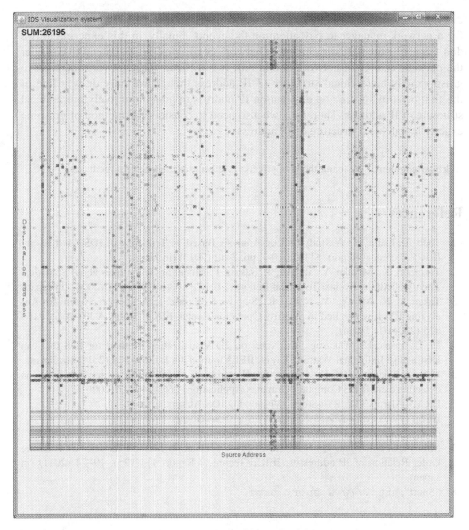

Fig. 13 Unequal Scaling visualization system with emphasizing anomalous events

It is expected that the emphasizing plots pays attention to the anomalous warning events to a network manager. In addition, animation mode enables us to analyze time variance of warning events.

5 Conclusion

This paper describes a study of IDS visualization system which emphasizes the anomalous warning events based on past tendency. The system is intended for reducing a load of a network manager. In the proposed method, we introduce calculation algorithm for deciding anomalous warning events. Since the anomalous warning events are plotted as emphasized shape on UnEqual Scaling Alert Matrix, it is easy for a network manager to pay attention the events. Finally, in this experiment, we don't consider the periodic change of network conditions. So we intend to consider the periodic change of network conditions. It is difficult for a network manager to grasp exactly hierarchical structure of IP address though we only draw a line of the first octet of source and destination IP address to grasp intuitively hierarchical structure of IP address. Therefore, we also need to consider the method of grasping exactly hierarchical structure of IP address.

Acknowledgment. We thank the anonymous reviewers for their helpful feedback.

References

1. Toda, T., Inaba, H.: A Study on Log Analysis Based on Tendency of IDS Alert Event. IEICE Technical Report, SITE2010-7, pp. 7–12 (2010) (in Japan)
2. Takemori, K., Miyake, Y., Tanaka, T., Sasase, I.: Modeling Techniques about Statistical Theory of Attack Events. Technical Report of IEICE 103(691), 20–27 (2004) (in Japan)
3. Takemori, K., Miyake, Y., Nakao, K., Sugaya, F., Sasase, I.: A Support System for Analyzing IDS Log Applied to Security Operation Center. IEICE Trans. A J87-A(6), 816–825 (2004) (in Japan)
4. Li, L., Inaba, H., Wakasugi, K.: Notes on 2D Visualization Method for IDS that can Distinguish Individual Warning Event. IIEEJ Journal 40(2), 369–376 (2011) (in Japan)
5. Itoh, T., Takakura, H., Koyamada, K.: Hierarchical visualization of network intrusion detection data. IEEE Computer Graphics Applications 26(2), 40–47 (2006)
6. I.R.V.I. Alarms: IDS RainStorm: Visualizing IDS Alarms, http://citeseerx.ist.psu.edu/viewdoc/download?doi=10.1.1.117.8777&rep=rep1&type=pdf
7. Mizoguchi, S., Inaba, H.: Proposal of 3D Visualization Method for IDS Considering Order Relation of IP addresses. IEICE Technical Report 111(125), 19–24 (2011) (in Japan)
8. "Snort", http://www.snort.org/

Affective Core-Banking Services for Microfinance

Insu Song and John Vong

Abstract. In this paper, we present a mobile phone-based banking system, called ACMB (Affective Cashless Mobile Banking), for microfinance. ACMB is designed to provide banking services to the world's unbanked population of 2.2 billion. To enable interoperability between various microfinance institutions over a heterogeneous network of mobile devices, cell-phone networks and internet services, we define MSDL (Microfinance Service Definition Language) based on WSDL (Web Service Definition Language). MSDL includes a binding for SMS for service queries and utilization. To ensure that the banking service provides an acceptable level of usability and user experience, ACMB incorporates a well-behaved service interface, which informs the design to create affective banking services based on human emotion models. ACMB was implemented on an Android tablet and evaluated with 147 participants, who performed 804 transactions and exchanged 2,412 SMS messages over a three hour testing period. The results suggest that an ACMB core-banking server on a low-cost mobile device can serve over 15,000 microfinance customers. Therefore, the systems appears to be suitable for most microfinance institutions.

Keywords: Mobile money, mobile payments, mobile banking, micro-enterprise, microfinance, cashless economy, affective computing.

1   Introduction

Banking is essential to any modern economy (Tan 2000), yet, 56% of the world's total population are still unbanked (Alberto Chaia 2010). The main reasons for this phenomena are the fixed costs of banking systems and the costs of running

Insu Song · John Vong
School of Business/IT, James Cook University Australia, Singapore Campus,
600 Upper Thompson Road, Singapore 574421
e-mail: {insu.song,john.vong}@jcu.edu.au

R. Lee (Ed.): *Computer and Information Science*, SCI 493, pp. 91–102.
DOI: 10.1007/978-3-319-00804-2_7   © Springer International Publishing Switzerland 2013

banking activities (e.g., credit scoring and lending) required by banks. A third (approximately 800 million people) of the unbanked population live in Asia. These people, who are in the lowest income category (i.e., living on under $5/day), simply cannot afford the high costs. It is only natural to ask why current banks do not have an efficient banking system capable of providing banking services to everyone. The fundamental problem is that the current commercial banking systems, policies, and operations are designed for the middle and upper classes. Therefore, a new approach is required.

In the last decades, various solutions have been suggested, such as solidarity lending and village banking, pioneered by Grameen Bank (Yunus 1999) in Bangladesh. Grameen Bank pioneered banking services for the poor. Their new banking model removed the operating costs and costs associated with risks. These approaches attempt to bridge the gap between the two incompatible partners: the unbanked population that is poor and commercial banks that are designed to serve the rich. However, these solutions are operated on manual systems that involve posting transactions from one ledger to another with human hands. Therefore, they do not scale well and are susceptible to fraud and other accounting irregularities.

The extensive penetration of cell phone networks and 2G devices in many of the world's most unbanked regions (global mobile phone subscriptions have reached 87% of the global population and 79% of developing world's population ((ITU) 2011)) suggests that mobile phone-based banking services could be the ideal solution for providing banking services to those regions. However, most of existing banking systems supporting mobile phone-based banking (various terms have been introduced in the past, such as "Mobile Phone Banking" or "Mobile Money Services" (Beshouri C. 2010; Donner 2008)), such as Sybase mCommerce 365 by SAP (Edgar 2009), and DELL Mobile Banking and Payments (DELL) are targeted at developed countries and, therefore, would incur additional overhead costs on top of the existing banking systems.

A new mobile core-banking system, called ACMB (Affective Cashless Mobile Banking), has been developed based on a cashless and wireless economic model to reduce operating costs. ACMB defines and exports its services through MSDL (Microfinance Service Definition Language), which is a subset of WSDL (Web Service Definition Language), to allow interoperability between various systems over heterogeneous mobile devices, cell-phone networks, and internet services. MSDL includes Short Message Service (SMS) binding to allow services to be discovered and utilized over SMS. To ensure the services are usable by 2.2 billion people, ACMB implements a Well-Behaved Service Interface (WBSI) based on human emotion models. A prototype of ACMB was implemented on an Android tablet and evaluated in a small artificial village comprising five mobile bank operators, five shops (sellers), seven companies (creditors), and 130 workers/buyers. The participants performed 804 transactions and sent 2,412 SMS messages over a three hour period. The results suggest that an ACMB system on a low-cost Android device can serve over 15,000 microfinance customers, making it large enough for most microfinance institutions. Section III reports on these performance results, as well as usability and user experience survey results.

2 Background

2.1 Empowering the Poor: Microfinance and Microenterprise

World Bank (World Bank 1999) defines a microenterprise as "an informal sector business with five workers or less, and fixed assets valued at less than US$10,000." This definition is similar to that used by USAID (Blayney and Otero 1985) and others (Ayyagari et al. 2007).

The research of de Mel et. al. (De Mel et al. 2010) reported that in low income countries, the self-employed make up around a third of the non-agricultural labor force and half or more of the informal sector,. The researchers also found that a substantial majority of these micro-entrepreneurs work by themselves and hire employees without salaries.

Tokman's (Tokman 2007) views are similar to those of the International Labour Organization (ILO). He suggested that the micro-entrepreneurs are actually being by-passed by the employment market. The "unemployed workers" start small enterprises to run, while hoping for paid employment. This view is in some way supported by the research of (De Soto 1989), who hypothesized that micro-entrepreneurs are in the informal sector because of three difficulties – lack of access to credit, inability to obtain approval for a business license, or being prevented from starting a business. Grameen Bank (Yunus 1999) in Bangladesh pioneered banking services which provide these services to the poor.

A study done by Vong et al. evaluated the impact of mobile money services on micro-entrepreneurs in rural Cambodia Developing countries (Vong et al. 2012). It showed that the financial access to electronic money has changed and improved livelihoods of rural Cambodia.

2.2 Mobile Banking

Various terms have been used for mobile banking, such as "Mobile Phone Banking" and "Mobile Money Services" (Beshouri C. 2010; Donner 2008). Examples of mobile banking systems include Sybase mCommerce 365 by SAP (Edgar 2009) and DELL Mobile Banking and Payments (DELL). In developed countries, much research has been conducted recently into how to make mobile phone-based banking accessible to people with disabilities (Pous et al. 2012) and security and trust issues (Zhou 2012), as well as public perceptions of and attitude toward such systems (Kim and Kang 2012).

3 Core Banking System for Microfinance

3.1 MSDL: Microfinance Service Definition Language

ACMB banking services are defined in MSDL (Microfinance Service Definition Language), which is a subset of WSDL (Web Service Definition Language). The distinctive feature is an SMS binding extension, called Structured SMS.

We represent a web service W as a structure:

$$W = (OP_i, OP_o, M, DB, PL)$$

where OP_i is a set of input operation types, OP_o is a set of output operation types, M is a set of message types, DB is a database definition (e.g., DTD speculations for XML views of relational databases), PL is a set of plans (or procedures), each of which achieves certain goals. OP_i, OP_o, and M can be obtained directly from the MSDL specification of banking services.

Let us represent each message type in M as follows:

$$\text{MessageTypeName}(Part_1,..., Part_n)$$

which is just an alternative representation of WSDL message definitions. We also represent each operation type in OP_i and OP_o as $OpName(m)$, where $OpName$ is the operation name of the operation and m is a message type.

3.2   Well-Behaved Banking Service Interface

According to the speech act theory of Searle (Searle 1985), each communication interaction is an attempt to perform an illocutionary act, such as a request or an assertion. Therefore, interactions can be classified into five types (illocutionary points) of illocutionary acts, but we find that mostly only four types (directive, assertive, commissive, declarative) are used in user-service communications. We define an operation-to-speech-act mapping function as follows:

$$MSmap : O \rightarrow U \times F \times Q$$

where O is a set of operations, U is a set of users, $F = \{$ directive, assertive, declarative, commissive$\}$ is a set of illocutionary points, Q is a set of XPath queries. A query is a specification for checking the achievement of an operation. $MSmap(op)$ returns a triple $(u; f; q)$ for an operation instance op. We now suppose that this mapping function is given in addition to the service definition W.

An instance op of incoming operation can have one of the following goal types, depending on the illocutionary point of the operation:

1. If f is a directive, the goal is op, meaning that the service achieves op. For example, Sell(Item('Book')).
2. If f is an assertive or declarative, the goal is $Bel(s, op)$, meaning that the service s believes op. For example, Details(PhoneNo('5555')).
3. If f is a commissive, the goal is $Bel(s, Int(u, op))$, meaning that the service s believes that the customer u intends to achieve op for the service, where $Bel(s, p)$ means s believes p, $Int(u, p)$ means u intends p, and thus $Bel(s, Int(u, p))$ means s believes that u intends p.

With this information on the illocutionary point of each message, we can now define how a bank service should interact with its users. In this paper, we consider two cases: directive one-way operations and commissive notification operations.

When users of a bank service know that the service has received a message containing the users' goal, and the service is responsible for achieving it, users expect the service to send them an acknowledgement and to inform them whether or not the goal has been achieved. The acknowledgement means that: (a) the message was received; (b) the goal will be achieved within a certain time limit; and (c) if the goal is achieved, the users will be informed. If the goal is a promised goal, acknowledgement is not necessary. In both cases, if the goal cannot be achieved within a certain time limit, the process must send a delay message asking users to wait for a certain period of time. These are just basic protocols that are expected by most human users. Similar protocols are also defined in an agent communication language called FIPA-ACL (FIPA [Fondation for Intelligent Physical Agents] 2002).

Let us call bank service interfaces conforming to the above descriptions well-behaved-interfaces. Figure 1 shows the state flow diagram for well-behaved-interfaces in a directive one-way operation. In the figure, upon receiving a directive operation, the service must respond within a certain response time rt_g with one of the following messages:

1. a confirming message $confirm_g$ or
2. a disconfirming message $disconfirm_g$ or
3. a delay message $delay_g(d)$ or
4. an acknowledgement message ack_g.

If the response is either confirming or $disconfirm_g$, the communication ends. But, if the response is $delay_g(d)$ or ack_g, the users expect one of the above messages (except ack_g) again within a certain goal achievement time, gt_g. If the following response is $delay_g(d)$, this time users expect one of the above messages (except ack_g) within d. This model is then used to deduce emotional states from the prospective events of banking services based on the work of the Ortony, Collins, and Clore's (OCC) (Ortony et al. 1990) cognitive appraisal theory of emotion, which is one of the most widely accepted emotion models. We apply the emotion defeasible rules (Song and Governatori 2006) to design the banking service operations to ensure they are well-behaved.

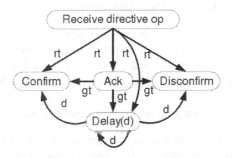

Fig. 1 Well behaved banking-service interface state flow diagrams for a directive one-way operation (e.g., SellItem). It is assumed that the response time, rt, is smaller (earlier) than the goal achievement time, gt: rt<<gt. For SMS binding, rt is about 4 seconds.

3.3   ACMB Prototype Design for Evaluation

In order to develop a banking system for microfinance, the following main design goals were set focusing on coverage, cost, and usability:

1. The system should be usable by any region with 2G cell phone networks, as 3G services are currently limited to metropolitan areas. This will ensure that we can cover 79% of the total developing world population.
2. The banking system must be affordable for most developing countries, such as Bangladesh.
3. The services should be usable (effective, efficient, easy to learn, helpful and convenient) for any mobile phone users.

The whole microfinance system must be easily operable by anyone, including mobile branches (e.g., mobile operators on motor bikes. The villagers, merchants, and microfinance bank officers should be able to perform the banking services (e.g., account creation, balance checking, lending, and money transfers) using any 2G mobiles phone capable of Short Message Service (SMS). For this, we have decided to implement the whole core banking server on a smart phone, which can automatically process SMS messages to provide banking services over SMS. The server then can be powered by batteries and carried on motor bikes by mobile bankers. The villagers, merchants, and microfinance bank officers can use the banking service by using any 2G mobiles phones capable of Short Messaging Service (SMS).

In this study, the mobile banking system comprises two types of devices: mobile bank servers operating on mobile devices capable of running a program for processing electronic messages, as well as sending and receiving electronic messages (e.g., SMS) and mobile stations (e.g., customers and mobile operators) that can send and receive SMS messages. In this paper, we describe a banking system having only one mobile bank server using SMS for communication. The mobile stations are used by bankers, buyers, or sellers. In this setting, we also assume that bankers operate the mobile bank server to manage accounts of all participants. This simple setting can provide the following basic trading services:

1. Bankers can create accounts for buyers and sellers.
2. Bankers can provide bank loans to buyers (lending service).
3. Sellers can sell goods to buyers and receive credits on their accounts.
4. Buyers can buy goods from sellers and credits can be deducted from their accounts.
5. Buyers and sellers can enquire about their credit balances.

To ensure the usability of the banking service, we reviewed the credit card transaction protocols (Bruce Binder 1996), where (1) merchants present invoices to buyers with the price, (2) buyers present a credit card for the payment, (3) merchants send the credit card information to a bank for authorization, and (4) products are then delivered to buyers. We adopted a similar protocol where only merchants send messages to banks for authorization. The same security and trust measures (e.g., the 24 hour grace period in the event of stolen or lost cards) can be used for our mobile banking system to provide the same or better level of service quality, security, and trust.

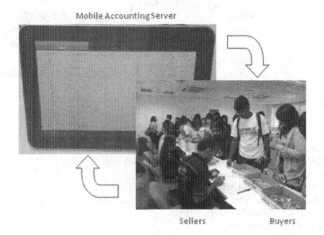

Fig. 2 The figure shows a seller sending an SMS message to the mobile bank server for a transaction. The server then sends authorization codes to the two parties (the buyer and seller) to complete the transaction.

Fig. 2 shows an overview of the mobile phone-based trading system that was used in our trial. The mobile bank server application runs on an Android tablet (SAMSUNG Galaxy Tablet 8.9). The server maintains an SQL database containing all account details of the participants including sellers, buyers, bankers, and lenders. The server can upload the account and transaction details to a remote backup system periodically to ensure the reliability of the service.

4   Implementation of the Prototype

From the inception of the idea, the design of the protocol and a prototype for the evaluation was completed within two months. The mobile bank server was tested on various Android phones (SAMSUNG Galaxy Note, SAMSUNG Galaxy Tab 8.9, and HTC Desire). Google Android SDK was used on Eclipse IDE using JAVA programming language with the target Android OS version 2.2. The users were able to receive authentication responses from the mobile bank server within ten seconds. This delay greatly influences service usability and depends on the cell-phone infrastructure.

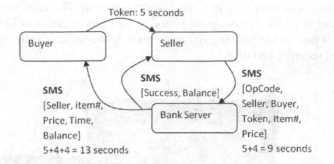

Fig. 3 A Buyer-Seller transaction. The whole transaction completes within 13 seconds.

```
IF op is CreateNewAccount:
  MSG = CreateNewAccount(PhoneNo, Details)
  SendSMS(PhoneNumber, MSG)
ELSEIF op is CreditAnAccount:
  MSG = CreditAnAccount(PhoneNo, CreditAmt)
  SendSMS(PhoneNumber, MSG)
ELSEIF op = SellItem:
  currentBalance =
    CurrentBalanceOfAccount(BuyerPhoneNo)
  IF currentBalance >= requestedAmount:
    MSG1, MSG2 =
      DebitAccount(BuyerPhoneNo, requestedAmount)
    SendSMS(SellerPhoneNo, MSG1)
    SendSMS(BuyerPhoneNo, MSG2)
  ELSE
    SendSMS(SellerPhoneNo, "Declined")
    SendSMS(BuyerPhoneNo,
      "Sorry you don't have enough credits in
        your account.")
ENDIF
```

Fig. 4 Mobile banking service algorithm used for the evaluation of MB4ME usability. SendSMS must be completed within the rt (response time) for Well-behavedness (see Fig. 1).

The credit card transaction protocol was implemented using a stateless rule-based decision tree. Figure 3 illustrates a transaction. The main algorithm of the test bank-server is shown in Figure 4. Solely for purposes of illustration, a simplified overview of the algorithm is shown below:

The algorithm describes how account creation, lending, and buying/selling banking activities are handled using a stateless algorithm. The server operates in a stateless mode except for the account balances: an incoming messages is processed only based on account numbers, roles, and available credits. Therefore, the system is highly reliable. When the server receives an SMS message, it checks if the sender exists in the account table using the sender's mobile phone number and the sender's role. Based on the role, the server processes the message content to create an account or to credit or deduct a certain amount to/from the buyer's or seller's account. Confirmation messages are also sent to the participants of the transactions. The participants will decide to trade physical goods according to the instructions sent by the bank server. Therefore, the trading system depends on the reliability of SMS messages. If SMS messages fail to be delivered within the response time, rt, the server must retry and inform system administrator to contact the participants. However, this would happen very rarely; therefore, its impact on the usability would be minimal.

5   Evaluation of ACMB Prototype

The trial of the mobile bank server was conducted during an event at a University in Singapore. A Polytechnic in Singapore organized student volunteers to participate in the trial. As an incentive, the top-five highest trading participants were rewarded with an iPod Nano. The students were encouraged to collect credit points by participating in events (as an analogy for lending or working for companies) and spend the credit points in a shop by purchasing items. We provided various food items, drinks, and souvenir items as shown in Fig. 2. We should note that the participants were not given any description of the banking system. They were only given very simple instructions to use their phone for transactions.

From 11am to 12pm, five bankers started to register the 130 RP students to create accounts for them. Upon registration, each participant was given an initial credit of $10. On average, it took less than three minutes to create an account for a new user and instruct them to use their phones for transactions. From 12pm to 2pm, the RP students participated in various events to earn credits. They were able to purchase items only between 12pm and 2pm.

There were two service interruptions due to: (1) a bug in the prototype, which was quickly fixed by the programmer on the fly; and (2) running out of the prepaid mobile account credit, which was solved by adding an additional pre-paid credit. At 2pm, the event was closed and a survey was conducted by RP staff. The survey consisted of eight multiple choice questions for assessing the usability of the mobile phone-based trading system and one open question for assessing user experiences.

5.1   Findings

The mobile bank server running on an Android tablet device was capable of handling 804 transactions within three hours. Over 1,608 SMS messages were sent from the mobile bank server (to banker/users, lender/users, and merchants/users) for the transactions. This means that the mobile banking server was able to support each transaction in more than 14 seconds. The server in fact handled over ten transactions at the same time at some points as there were five sellers and seven lenders who were working at the same time. Given that the theoretical limit of a transaction delay (assuming the average SMS delivery time requires four seconds) is eight seconds, the prototype mobile banking server running on a battery powered smart phone showed an impressive performance. This clearly demonstrates that the mobile banking system can deliver microfinance services effectively and efficiently.

Table 1 Survey Questions and Results

Q#	Quetions	Average	Std Dev
Q1	The mobile trading system is easy to use	1.84[*]	0.66
Q2	I think that I would need the help of other people to be able to use this mobile	2.43	0.86
Q3	The steps required to perform mobile trading were easy to follow.	1.80*	0.61
Q4	I was able learn to use this system very quickly.	1.69*	0.62
Q5	Most people would learn to use this system very quickly.	1.84*	0.65
Q6	I am confident that I would remember how to use the system again.	1.81*	0.65
Q7	I would like to use this system again.	2.00*	0.72
Q8	I would recommend this system to others.	1.97*	0.68

The answers were based on 4 point Likert scale: 1 = Strongly Agree, 2 =Agree, 3 = Disagree, 4 = Strongly Disagree.

Among the 130 participants, 128 completed the survey: a 98% re rate. The survey results suggest that most participants agreed that the mobile trading system was easy to use, easy to learn, and memorable. Most participants also agreed that they would use the trading system again and would recommend the system to others. The results were statistically significant ($p<0.01$). Table 1 illustrates the survey results. Those with statistical significance (significantly different from 2.5) are highlighted.

The mobile bank server and the trial were initially designed for 50 target users, but due to popular responses from participants, we had over 130 participants. Because of this, there were long queues at the registration booths, shops, and lenders (activity managers of the companies) as there were only five bankers for registration, five shopkeepers, and seven activity managers. Many participants were frustrated by the long queues, and this was reflected in the Question 7 and on the open ended question.

6   Conclusion

This study strongly suggests that it is feasible to provide an affordable banking system for microfinance utilizing only 2G cell phone networks. This is significant since the current availability of 3G networks is limited in many of the developing countries. The total cost of ownership of the banking system could go as low as US$100. The system would be able to support about 2,000 transactions (assuming 10 seconds per transaction limited by the current SMS services).

According to this study, the average users would feel comfortable using mobile phone-based transactions, thereby lessening doubts about whether villagers would use such a system. The mobile banking server was able to support the credit-card-like transaction protocols ensuring that users would experience the same level of security, trust, and service quality of credit card services. In comparison with a mobile money service called WING Money launched in Cambodia in 2008, ((WING) 2009), the protocol is much simpler, eliminating the need for a money agent. The main reason is because the mobile banking system can support not only lending, but also merchant transactions, thereby creating a cashless transactions environment.

The mobile bank server can also efficiently generate daily transaction reports and incorporate other charges such as GST. Therefore, governments would welcome the use of the mobile banking server, since it can be used to collect taxes and keep track of economic activity among a country's unbanked population, which would otherwise be impossible.

References

(ITU) ITU. ITU World Telecommunication/ICT International Telecommunications Union (2011)
(WING) WMC. Social Impact Report. WING Money Cambodia (2010),
        http://www.wingmoney.com/fileadmin/templates/main/
        images/Social_Impact_Report_2009.pdf
Alberto Chaia, A.D., Goland, T., Mayasudhakar, M.J., Morduch, J., Schiff, R.: Half The World Is Unbanked. Financial Access Initiative (2010),
        http://financialaccess.org/sites/default/files/publica
        tions/B25_Half_the_World_is_Unbanked%282011-01-21%29-
        1.pdf (accessed May 17, 2010)
Ayyagari, M., Beck, T., Demirguc-Kunt, A.: Small and medium enterprises across the globe. Small Business Economics 29(4), 415–434 (2007)
Beshouri, C., Chaia, A., Cober, B., Gravrak, J.: Banking on Mobile to Deliver Financial Services to the Poor. In: Mertz, R. (ed.) Global Financial Inclusion, pp. 24–31. McKinsey & Company, USA (2010)
Blayney, R.G., Otero, M.: Small and Micro-Enterprises: Contributions to Development and Future Directions for AID's Support. United States. Agency for International Development (1985)
Bruce Binder, J.G., Hart, A., Kiser, S., Klebe, S., Knox, O., Lee, T., Lindenberg, B., Lum, J., Melton, B., Paredes, D., Chintamaneni, P., Silverman, F., Wilson, B., Wong, G., Wu, W., Zalewski, M.: CyberCash Credit Card Protocol Version 0.8. Network Working Group (1996),
        http://tools.ietf.org/html/rfc1898#section-1.3 (2012)

De Mel, S., McKenzie, D., Woodruff, C.: Who are the Microenterprise Owners? Evidence from Sri Lanka on Tokman versus De Soto. In: International Differences in Entrepreneurship, pp. 63–87. University of Chicago Press (2010)

De Soto, H.: The Other Path: The Invisible Revolution in the Third World. Harper and Row, New York (1989)

DELL. Mobile Banking and Payments. DELL (2012),
    `http://i.dell.com/sites/doccontent/business/solutions/`
    `engineering-docs/en/Documents/mobile-banking-`
    `payments.pdf`

Donner, J., Tellez, C.A.: Mobile Banking and Economic Development: Linking Adoption, Impact, and Use. Asian Journal of Communication 18(4), 318–332 (2008)

Edgar, D.C.: Realizing the Full Potential of Mobile Commerce Orchestrating Mobile Payments and Money Transfers. Edgar, Dunn & Company (2012),
    `http://www.sybase.com/mCommercewhitepaper`

FIPA [Fondation for Intelligent Physical Agents], FIPA ACL message structure specification (2002)

Kim, J.B., Kang, S.: A study on the factors affecting the intention to use smartphone banking: The differences between the transactions of account check and account transfer. International Journal of Multimedia and Ubiquitous Engineering 7(3), 87–96 (2012)

Ortony, A., Clore, G.L., Collins, A.: The cognitive structure of emotions. Cambridge university press (1990)

Pous, M., Serra-Vallmitjana, C., Giménez, R., Torrent-Moreno, M., Boix, D.: Enhancing accessibility: Mobile to ATM case study, pp. 404–408 (2012)

Searle, J.R.: Expression and meaning: Studies in the theory of speech acts. Cambridge University Press (1985)

Song, I., Governatori, G.: Affective web service design. In: Yang, Q., Webb, G. (eds.) PRICAI 2006. LNCS (LNAI), vol. 4099, pp. 71–80. Springer, Heidelberg (2006)

Tan, M., Teo, T.S.H.: Factors influencing the adoption of internet banking. Journal of the Association for Information Systems 1(5), 1–42 (2000)

Tokman, V.E.: Modernizing the informal sector. UN/DESA Working Paper (42), 1–13 (2007)

Vong, J., Fang, J., Song, I.: Delivering financial services through mobile phone technology: a pilot study on impact of mobile money service on micro–entrepreneurs in rural Cambodia. International Journal of Information Systems and Change Management 6(2), 177–186 (2012)

World Bank, The World Bank and Microenterprise Finance: From Concept to Practice. Operations Evaluation Department, World Bank (1999)

Yunus, M.: Banker to the Poor: Micro-Lending and the Battle Against World Poverty. PublicAffairs, New York (1999)

Zhou, T.: Understanding users' initial trust in mobile banking: An elaboration likelihood perspective. Computers in Human Behavior 28(4), 1518–1525 (2012)

Shadow Detection Method Based on Shadow Model with Normalized Vector Distance and Edge

Shuya Ishida, Shinji Fukui, Yuji Iwahori, M.K. Bhuyan, and Robert J. Woodham

Abstract. An object detection method needs a shadow detection because shadows often have a harmful effect on the result. Shadow detection methods based on shadow models are proposed. A new shadow model is proposed in this paper. The proposed model is constructed by the differences of UV components of YUV color space between the background image and the observed image, Normalized Vector Distance and edge information. The difference of Y component is less suitable for the model because it varies considerably with location. The proposed model includes Normalized Vector Distance instead of Y component. It is a robust feature to illumination changes and can remove a shadow effect in part. The proposed method can obtain shadow regions more accurately by including Normalized Vector Distance in the shadow model. Results are demonstrated by the experiments using the real videos.

Keywords: Shadow Detection, Shadow Model, Normalized Vector Distance, Edge, Gaussian Mixture Model.

Shuya Ishida · Yuji Iwahori
Chubu University, Kasugai, Aichi, 487-8501 Japan
e-mail: ishida@cvl.cs.chubu.ac.jp, iwahori@cs.chubu.ac.jp

Shinji Fukui
Aichi University of Education, Kariya, Aichi, 448-8542 Japan
e-mail: sfukui@auecc.aichi-edu.ac.jp

M.K. Bhuyan
Indian Institute of Technology Guwahati, Guwahati - 781039, India
e-mail: mkb@iitg.ernet.in

Robert J. Woodham
University of British Columbia, Vancouver, B.C. Canada V6T 1Z4
e-mail: woodham@cs.ubc.ca

R. Lee (Ed.): *Computer and Information Science*, SCI 493, pp. 103–113.
DOI: 10.1007/978-3-319-00804-2_8 © Springer International Publishing Switzerland 2013

1 Introduction

In many fields of computer vision, a method for detecting moving objects is used as a preprocessing. Many methods for detecting moving objects have the problem of detecting a shadow as an object. Shadows often have a harmful effect on the result. There are cases that some objects are extracted as one object because of shadows, for example. Methods for detecting shadows and for removing them have been proposed[1, 2, 3, 4, 5, 6, 7, 8].

Methods using multiple cameras have been proposed[1, 2]. These methods cannot be applied to an image obtained through one camera. Methods using a camera have a wider application than those using multiple cameras. In this paper, a method using one camera is treated.

The methods[3, 4] convert color space to detect shadows. It is difficult for them to detect all shadows stably in any conditions. On the other hand, the methods[5, 6, 7, 8] use the shadow models. The methods[5, 6] model shadows by the mixture distribution models. These methods should determine the numbers of the distributions in advance. It is difficult to determine the proper number of the distributions. To solve the problem, the method based on the shadow model constructed by the nonparametric Bayesian scheme is proposed[7].

In the case of using a shadow model, it should be updated because a method using a model cannot detect shadows which are not included in the learning data of the shadow model. The model should be updated by only the shadow data because the suitable shadow model cannot be constructed and the method sometimes fails in detecting shadows when the learning data include the data which are not shadow data. The method for detecting shadows and for updating the shadow model is proposed[8]. The method[8] can select the learning data. Obtaining more accurate results remains as a future work.

This paper proposes a new shadow model to extract shadows more accurately than the previous approaches. The proposed method constructs the shadow model by color and edge information of shadow regions. The color information is used after converting from RGB to YUV. The proposed approach uses Normalized Vector Distance (ND)[9] instead of Y component. ND is a robust feature to illumination changes and can remove shadow effects in part. The result detected by the shadow model is improved by using the method based on object regions and the color segmentation method. The improved result includes shadows which are not included in the learning data. They are used for updating the shadow model. Results are demonstrated by the experiments using the real videos.

2 Shadow Model

The proposed method constructs the shadow model by color and edge information.

2.1 Shadow Model by Color Information

The proposed method uses U and V components of YUV color space and uses a background image. The differences of U and V components of each pixel in shadow regions between the background image and a target image become small. On the other hand, those in a moving object region become large. The method uses the differences of UV components as the observed data.

The method[8] uses the difference of Y component to model shadows. It is less suitable to use it as the data for the shadow model because the difference varies considerably with location. The proposed method uses ND instead of Y component. ND is a robust feature to illumination changes and can remove shadow effects in part. ND is calculated by Eq. (1) after dividing an image into small blocks.

$$ND(x) = \left| \frac{D_i(x)}{|\mathbf{i}(x)|} - \frac{D_b(x)}{|\mathbf{b}(x)|} \right| \tag{1}$$

where $D_i(x)$ and $D_b(x)$ mean the irradiance data at a pixel x in the observed image and the background image respectively, $\mathbf{i}(x)$ means an irradiance vector for a block of the observed image which includes x and $\mathbf{b}(x)$ means that for the block of the background image. The proposed method uses Y component at x for the data.

Let a datum which consists of UV component and ND at a pixel x be represented by $D(x)$. $D(x)$ in a shadow region is used to make the shadow model. After the frequency distribution of the data is obtained, the probability of occurrence for the frequency distribution is approximated by the Gaussian mixture distribution. It is used as the shadow model by the color information. The shadow model by the color information is represented by Eq. (2).

$$P_C(S(x)|\alpha, \theta, D(x)) = \sum_{k=1}^{K} \alpha_k \mathcal{N}(D(x); \theta_k) \tag{2}$$

$$\alpha = \{\alpha_1, \cdots, \alpha_K\}, \quad \theta = \{\theta_1, \cdots, \theta_K\}$$

where $S(x)$ means the state where x exists in a shadow region, K means the number of the distribution, $\mathcal{N}(\cdot; \theta_k)$ means k-th Gaussian distribution. α_k and θ_k are the mixture ratio and the parameters of k-th distribution respectively.

The parameters of the distribution are estimated by a nonparametric Bayesian scheme. The proposed method uses the Dirichlet Process EM(DPEM) algorithm[10] to estimate the parameters of the distribution. It is based on the EM algorithm and can be implemented easily. The algorithm can calculate not only the parameters of the mixture distribution but also the mixture ratios. When the parameters are estimated, the DPEM algorithm uses a sufficiently large number of distributions to truncate the approximation. The proposed method sets the number for truncation to 20. After estimating the mixture ratios, the number of the distribution for the shadow model is determined according to the mixture ratios. The model is obtained by removing distributions with little mixture ratios from the estimated mixture distribution.

2.2 Shadow Model by Edge Information

The difference of the edge information of each pixel in shadow regions between the background image and a target image becomes small. On the other hand, that in a moving object region becomes large. A method using edge information for shadow model has been proposed[11]. The proposed method also uses edge information for the shadow model. The shadow model based on edge information is represented by the following equation.

$$P_E(S(x)|E_m(x),E_d(x)) = \lambda_1 \exp(-E_m(x)/w_1) + (1-\lambda_1)\exp(-E_d(x)/w_2) \quad (3)$$

where $E_m(x)$ and $E_d(x)$ mean the difference of the edge magnitude and that of the edge gradient direction between the current image and the background image respectively, and λ_1 $(0 \le \lambda_1 \le 1)$ means the weight. w_1 and w_2 mean the parameters which can tune the variances of the exponentials.

3 Shadow Detection

After constructing the shadow model, shadows are detected by using the model. The shadow detection process is applied to pixels extracted by the background subtraction. The detection process is as follows: At first, the probability that a pixel exists in a shadow region is calculated. Next, the shadow regions and the object regions are separated through the threshold processing. At last, the result is refined by the method based on the object regions obtained through the threshold processing and the method based on the color segmentation[12]. The detection process and the process for refining the result are explained in the following.

3.1 Shadow Detection by Using Shadow Model

The probability that x belongs to the shadow region is calculated by Eq. (4).

$$P_1(S(x)|\alpha,\theta,E_m(x),E_d(x),D(x)) = P_C(S(x)|\alpha,\theta,D(x)) \times P_E(S(x)|E_m(x),E_d(x))$$
$$(4)$$

$P_1(S(x)|\alpha,\theta,E_m(x),E_d(x),D(x))$ becomes large when x exists in a shadow region. x is regarded as the pixel in the shadow region when $P_1(S(x)|\alpha,\theta,E_m(x),E_d(x),D(x))$ is larger than a threshold. Otherwise, x is regarded as the pixel in the object region.

3.2 Refining Shadow Detection Result

Some misjudged pixels exist in the result through the above process. After the threshold processing, the process for refining the result is applied to it. The refinement process consists of two parts. At the first step, the result is refined by using the object regions obtained by the above process. Next, the color segmentation[12] is used for refining it more. These steps are described in the following.

Foreground pixels misjudged as shadow pixels exist in the object regions extracted through the above process. The misjudged pixels are reduced by using the information of the object regions. Let the labeled object region be represented by $R_{(i)}$ $(i = 1, \cdots, N)$, where N means the number of the object regions. At this time, the small regions are removed as noises. The probability that x does not exist in the object region is calculated by the following equation.

$$P_R(S(x_{(i)})|\theta_{R_{(i)}}) = 1 - \exp\left(-c\overline{x_{(i)}}^\top \Sigma_{R_{(i)}}^{-1} \overline{x_{(i)}}\right) \tag{5}$$

where $x_{(i)}$ means a pixel of which the nearest region is $R_{(i)}$, $\theta_{R_{(i)}} = \{\mu_{R_{(i)}}, \Sigma_{R_{(i)}}\}$, $\mu_{R_{(i)}}$ means the center position of $R_{(i)}$, $\Sigma_{R_{(i)}}$ means the variance-covariance matrix calculated by the positions of the pixels in $R_{(i)}$, $\overline{x_{(i)}} = x_{(i)} - \mu_{R_{(i)}}$.

The probability that x exists in the shadow region and does not exist in the object region is defined by Eq. (6). Each pixel is judged as the shadow or the object pixel again through the threshold processing.

$$P_2(S(x_{(i)})|\alpha, \theta, \theta_{R_{(i)}}, E_m(x_{(i)}), E_d(x_{(i)}), D(x_{(i)})) = \lambda_2 P_R(S(x_{(i)})|\theta_{R_{(i)}}) \times$$
$$P_1(S(x_{(i)})|\alpha, \theta, E_m(x_{(i)}), E_d(x_{(i)}), D(x_{(i)})) + (1 - \lambda_2) P_R(S(x_{(i)})|\theta_{R_{(i)}}) \tag{6}$$

After applying the above process, the observed image is segmented by color and it is judged whether each region belongs to the shadow region. The Mean-Shift segmentation[12] is used for color segmentation. A region where more than 60% pixels in the region are regarded as pixels in the shadow region by the above process is regarded as the shadow region. It is determined that all pixels in the regions exist in the shadow region. Otherwise, it is regarded as the object region and all pixels in the region are regarded as pixels in the object region.

4 Updating Shadow Model by Color Information

The proposed method obtains manually the data in the shadow regions to learn the initial shadow model by the color information. The shadow model is updated in order to detect shadows which are not included in the learning data. The shadow regions detected by the process explained in the previous section are used for updating the model. The process for updating the model is as follows:

STEP 1 Obtaining the learning data
STEP 2 Obtaining the initial values for DPEM algorithm
STEP 3 Updating the shadow model by DPEM algorithm

STEP 1 and STEP 2 are described in the following.

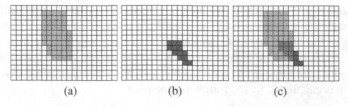

(a)   (b)   (c)

Fig. 1 Process for Updating Shadow Data: (a) Pixels Having Shadow Data at $t-1$, (b) Pixels Obtained as Current Shadow Resion, and (c) Pixels Having Shadow Data at t

4.1 Obtaining Learning Data

The proposed method stores the data of pixels which were detected as the shadow pixels at the past frames. When all data of the past frames are stored, a great deal of time is required to handle the DPEM algorithm because the number of data becomes enormous. Furthermore, the impact of shadow data which are obtained at the current frame becomes small. The proposed method selects data. The method for obtaining the learning data is described in the following.

First, misjudged pixels in the object region are removed from the current shadow data. The data of pixels extracted as the object pixels are used for removing misjudged pixels. The following process is applied to each object region labeled through the process described in the section 3.2. The variance-covariance matrix for the coordinates of the pixels in the object region is calculated and 95% probability ellipse is obtained. The data of the pixels which exist inside the ellipse are regarded as pixels in the object region and they are removed.

The obtained data through the above processes are added to the learning data. At this time, the data of pixels which were regarded as the shadow region at the past frames are overwritten. Fig. 1 shows this process. The data of the red pixels in Fig. 1-(c) are added to the shadow data and those of the blue pixels are overwritten by the current shadow data. This process prevents the method from handling the enormous data.

4.2 Obtaining Initial Parameters for DPEM Algorithm

Accuracy of the estimated result by the DPEM algorithm depends on initial values of parameters. The following process is applied to give better initial parameters. Parameters which should be given are mean vector and variance-covariance matrix for each Gaussian distribution. A sufficiently large number of distributions for truncation is also needed.

The number of the distributions and the initial parameters of the mixture distribution are determined by the following process.

STEP I   Obtaining data which are not included in the learning data
STEP II   Applying the DPEM algorithm to the data obtained by **STEP I**

(a) (b) (c)

Fig. 2 Process for Detecting Data not Included in Previous Learning Data: (a) Result by Threshold Processing, (b) Result by Segmentation Process, and (c) Data used at **STEP2**

At **STEP I**, the data which are not included in the learning data are obtained. The pixels of which the judgments are corrected to the shadow region from the object region are obtained. Fig. 2 shows the process for **STEP I**. Blue pixel in the figure means it is regarded as a pixel in the object region and red one means it is regarded as a pixel in the shadow region. The red pixels in Fig. 2-(c) are used for the learning data at **STEP II**.

At **STEP II**, the DPEM algorithm is applied to the data obtained through **STEP I**. The sufficiently large number of distributions for truncation for this process is set to 20. The initial mean vector for each distribution is set to a datum obtained randomly from the data. The initial variance-covariance matrix is set to the variance-covariance matrix which is calculated from the data. After the DPEM algorithm is applied, the number of distributions for the data is determined according to the mixture ratios.

The parameters for the current shadow model and those obtained through the above process are given for the initial parameters for the DPEM algorithm. The truncation number is obtained by adding the number obtained through **STEP II** to the number of the distributions for the current shadow model. Better initial parameters and smaller truncation number can be given by these processes.

5 Experiments

The experiments using the real images were done to confirm the effectiveness of the proposed method. The size of each frame is 720×480 pixels and each pixel has 8bit color value for each RGB component. Four scenes (Scene 1 - Scene 4) were used in the experiments. All scenes were taken outdoors.

The examples of input images used for the experiments and the results of the shadow detection are shown in Fig. 3 - Fig. 6. The blue pixels in the result images mean the object pixels and the red pixels mean the shadow pixels. The results show that the proposed method can detect most shadows. Many pixels in the regions of the shoes and the heads are misjudged as those in the shadow. The colors of them

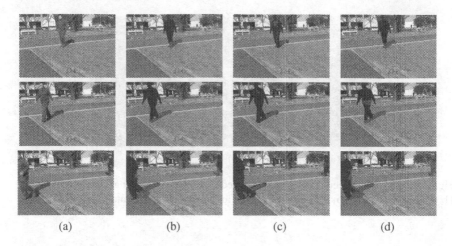

(a)   (b)   (c)   (d)

Fig. 3 Input Images and Results for Scene 1: (a) Input Images, (b) Results of Proposed Method, (c) Results of Method[5], and (d) Results of Method [8]

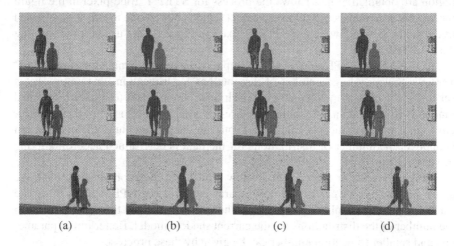

(a)   (b)   (c)   (d)

Fig. 4 Input Images and Results for Scene 2: (a) Input Images, (b) Results of Proposed Method, (c) Results of Method[5], and (d) Results of Method [8]

are similar to those of the shadow. The proposed method uses color information to construct the shadow model. It sometimes misjudges in object regions similar to colors of shadows. The experimental results by the methods[5, 8] are also shown in the figures to compare them with the proposed method. It is shown that the proposed method can get better results than the methods. The proposed method uses ND to construct the shadow model. The comparison of the results of our method and the method[8] shows that using ND instead of Y component causes constructing better shadow model and obtaining better result.

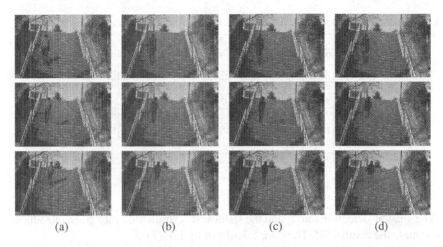

Fig. 5 Input Images and Results for Scene 2: (a) Input Images, (b) Results of Proposed Method, (c) Results of Method[5], and (d) Results of Method [8]

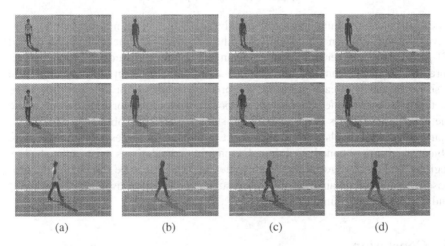

Fig. 6 Input Images and Results for Scene 4: (a) Input Images, (b) Results of Proposed Method, (c) Results of Method[5], and (d) Results of Method [8]

The processing times of the proposed method for Scene 1 were measured. The average time for the shadow detection process was 4.524 sec/frame. The process for the Mean-Shift segmentation took 4.432 seconds of 4.524 seconds. The time for the process for updating the model was 3.477 seconds. The processes of the Mean-Shift segmentation and the DPEM algorithm are very heavy loaded process. The devices, such as handling the updating process on other computer and using other methods instead of them, are required in order to detect a shadow in real time.

Table 1 Shadow Detection Rates (η) and Shadow Discrimination Rates (ξ)

		Proposed Method			Method [5]			Method [8]		
η[%]	Scene 1	97.29	96.00	96.57	60.72	57.85	51.85	97.29	95.94	96.37
	Scene 2	97.57	97.76	90.24	80.78	80.38	95.44	97.63	97.52	90.24
	Scene 3	90.33	96.78	96.24	78.30	60.69	73.41	80.84	88.27	81.98
	Scene 4	95.39	95.48	96.10	46.58	44.75	56.05	95.34	95.49	96.10
ξ[%]	Scene 1	91.29	94.51	84.85	90.19	97.78	96.89	90.68	87.07	66.87
	Scene 2	100.0	99.88	100.0	82.50	90.01	98.56	96.05	89.84	100.0
	Scene 3	99.59	98.83	91.68	95.68	95.88	97.61	99.85	75.83	88.32
	Scene 4	98.17	97.51	89.56	81.45	82.37	80.87	94.94	94.07	70.61

The shadow detection rate η and the shadow discrimination rate ξ are introduced to evaluate the results[13]. They are calculated by Eq. (7).

$$\eta = \frac{TP_s}{TP_s + FN_s}, \quad \xi = \frac{\overline{TP_f}}{TP_f + FN_f} \tag{7}$$

where TP means the number of true positives, FN means the number of false negatives, the subscription s means shadow, the subscription f means foreground and $\overline{TP_f}$ means the number subtracting the number of points misjudged as shadows on foreground objects from the correct number of points of foreground objects. Table 1 shows η and ξ of each scene for the proposed method, the method[5] and the method[8]. Some of the values of the proposed method are worse than those of the method[8]. η becomes larger when more shadow pixels are obtained even though many misjudged pixels exist in the object regions. ξ becomes larger when more object pixels are obtained even though many misjudged pixels exist in the shadow regions. The result is good when both values are high. It can be shown that the proposed method can obtain more suitable results than the previous method[8] when η and ξ are compared simultaneously.

6 Conclusion

This paper proposed a shadow model for a shadow detection method. The proposed shadow model is constructed by color and edge information. Normalized Vector Distance is used as one of the color information for constructing the shadow model. The shadow model including it causes obtaining the shadow region more accurately. The proposed method updates the shadow model to detect shadows which are not included in the learning data. The experimental results show that the proposed approach can obtain shadow regions more robustly than the previous approaches.

Future work includes obtaining initial shadow data automatically. Processing the whole process faster is also needed.

Acknowledgment. Fukui's research is supported by JSPS Grant-in-Aid for Young Scientists (B) (23700199). Iwahori's research is supported by JSPS Grant-in-Aid for Scientific Research (C) (23500228) and Chubu University Grant. Woodham's research is supported by the Natural Sciences and Engineering Research Council (NSERC).

References

1. Madsen, C.B., et al.: Shadow Detection in Dynamic Scenes Using Dense Stereo Information and an Outdoor Illumination Model. In: Proc. of DAGM 2009 Workshop on Dynamic 3D Imaging, pp. 110–125 (2009)
2. Iwama, H., Makihara, Y., Yagi, Y.: Foreground and Shadow Segmentation Based on a Homography-Correspondence Pair. In: Kimmel, R., Klette, R., Sugimoto, A. (eds.) ACCV 2010, Part IV. LNCS, vol. 6495, pp. 702–715. Springer, Heidelberg (2011)
3. Grana, C., et al.: Improving shadow suppression in moving object detection with hsv-color information. In: Proc. IEEE Intelligent Transportation Systems Conf., pp. 334–339 (2001)
4. Blauensteiner, P., et al.: On colour spaces for change detection and shadow suppression. In: Proc. 11th Computer Vision Winter Workshop, Telc, Czech Republic, pp. 87–92 (2006)
5. Tanaka, T., Shimada, A., Arita, D., Taniguchi, R.-I.: Non-parametric background and shadow modeling for object detection. In: Yagi, Y., Kang, S.B., Kweon, I.S., Zha, H. (eds.) ACCV 2007, Part I. LNCS, vol. 4843, pp. 159–168. Springer, Heidelberg (2007)
6. Wang, Y., et al.: Detecting shadows of moving vehicles based on hmm. In: ICPR 2008, pp. 1–4 (2008)
7. Kurahashi, W., Fukui, S., Iwahori, Y., Woodham, R.J.: Shadow Detection Method Based on Dirichlet Process Mixture Model. In: Setchi, R., Jordanov, I., Howlett, R.J., Jain, L.C. (eds.) KES 2010, Part III. LNCS, vol. 6278, pp. 89–96. Springer, Heidelberg (2010)
8. Fukui, S., et al.: A Method of Learning Data Selection for Updating Shadow Model with High Accuracy. In: Frontiers in Artificial Intelligence and Applications, Advances in Knowledge-Based and Intelligent Info. and Eng. Sys., vol. 243, pp. 1758–1767 (2012)
9. Nagaya, S., et al.: Moving Object Detection by Time-Correlation-Based Background Judgement Method. Trans. of IEICE J79-D-II(4), 568–576 (1996) (in Japanese)
10. Kimura, T., et al.: Semi-supervised learning scheme using dirichlet process em-algorithm. IEICE Tech. Rep. 108(484), 77–82 (2009)
11. Joshi, A.J., et al.: Moving Shadow Detection with Low- and Mid-Level Reasoning. In: IEEE International Conference on Robotics and Automation 2007, pp. 4827–4832 (2007)
12. Comaniciu, D., et al.: Mean Shift Analysis and Applications. In: Proc. of the International Conf. on Computer Vision, vol. 2, pp. 1197–1203 (1999)
13. Prati, A., et al.: Algorithms and Evaluation. IEEE Trans. on PAMI 25(7), 918–923 (2003)

Defect Classification of Electronic Board Using Multiple Classifiers and Grid Search of SVM Parameters

Takuya Nakagawa, Yuji Iwahori, and M.K. Bhuyan

Abstract. This paper proposes a new method to improve the classification accuracy by multiple classes classification using multiple SVM. The proposed approach classifies the true and pseudo defects by adding features to decrease the incorrect classification. This approach consists of two steps. First, the features are extracted from the defect candidate region after extracting the difference between the test image and the reference image. Here, candidate extraction is carefully extracted with high accuracy and the useful combination of features is determined using the feature selection. Second, selected features are learned with multiple SVM and classified into the class. When the result has the multiple same voting counts to the same class, the judgment is treated as the difficult class for the classification. It is shown that the proposed approach gives efficient classification with the higher classification accuracy than the previous approaches through the real experiment.

Keywords: Support Vector Machine, Multiple Class Classification, Extraction of Defect Candidate, Defect Classification, Grid Search.

1 Introduction

Basically, Printed Circuit Board (PCB) is a piece of phenolic or glass epoxy board with copper clad on one or both sides. The portion of copper that are not needed are etched off, leaving 'printed' circuits which connects the components. It is used to mechanically support and electrically connect electronic components using conductive pathways, or traces, etched from copper sheets and laminated onto a

Takuya Nakagawa · Yuji Iwahori
Graduate School of Engineering, Chubu University, 1200 Matsumoto-cho,
Kasugai 487-8501, Japan
e-mail: tnakagwa@cvl.cs.chubu.ac.jp, iwahori@cs.chubu.ac.jp

M.K. Bhuyan
Dept. of Electronics & Electrical Engg. IIT Guwahati, Guwahati - 781039, India
e-mail: mkb@iitg.ernet.in

R. Lee (Ed.): *Computer and Information Science*, SCI 493, pp. 115–127.
DOI: 10.1007/978-3-319-00804-2_9   © Springer International Publishing Switzerland 2013

non-conductive substrate. PCBs are rugged, inexpensive and highly reliable and so it is used in virtually all but the simplest commercially produced electronic devices.

It has become one of the basic components of electronic devices. It provides the electrical connections between the electronic or IC components mounted on it. In recent years, the demand of electronic devices with more compact design and more sophisticated functions has forced the PCBs to become smaller and denser with circuits and components. As it is crucial part of electronic device it needs to be properly investigated before get launched. Automatic inspection systems are used for this purpose but due to more complexity in circuits, PCB inspections are now more problematic. This problem leads to new challenges in developing advanced automatic visual inspection systems for PCB.

Automatic Optical Inspection (AOI) has been commonly used to inspect defects in Printed circuit board during the manufacturing process. An AOI system generally uses methods which detects the defects by scanning the PCB board and analyzing it. AOI uses methods like Local Feature matching, image Skeletonization and morphological image comparison to detect defects and has been very successful in detecting defects in most of the cases but production problems like oxidation, dust, contamination and poor reflecting materials leads to most inevitable false alarms. To reduce the false alarms is the concern of this paper.

Previous approach Tanaka *et al.* [1] classifies the defects using neural network and Rau *et al.* [2] proposes a method to classify the defects using the intensity at the pixels around the defects region. These approaches classify the defect under the condition that kinds of the defects are previously known. There are some defects whose recognitions are difficult even with the visual inspection. These defects cause the problem. The problem includes the case of misjudgment where a true defect is recognized as a pseudo defect and it is included in the products as a result. Kondo *et al.* [3] has been proposed for the distinction of defect classification by determining the features at random. Kondo *et al.* [3] classifies the kinds of defect with selecting the appropriate features with classifiers, but there are still incorrect classification cases where a true defect is classified into a pseudo defect.

Approaches to extract the defect candidate region are proposed in [4, 5, 6]. Onishi *et al.* [4] prepares two images of test image and reference image of mask pattern and takes difference image by logical AND. Maeda *et al.* [5] and Numada *et al.* [6] propose IR image matching and Mahalanobis distance, respectively.

Kondo *et al.* [7] selects features at random and determines the combination of features with bagging process to classify the defects. Other classification approaches include Wakabayashi *et al.* [8] using PCA, Ishii *et al.* [9] using variance inside and outside classes, Amabe *et al.* [10] using Genetic Algorithm, Roh *et al.* [11] using Neural Network. Another approach to remove incorrect classification of true defect is proposed in Iwahori *et al.* [12, 13], where histogram for each defect and evaluating equation are introduced.

The approach discriminates the true defect and pseudo defect and improves the accuracy of classification by introducing multiple classes for the real defect dataset. Experiments give the usefulness of the approach through the evaluation of defect classification.

2 Defect Classification Using Multiple Classifiers and Grid Search of Parameters

Defect candidate region is extracted from both test image and reference image. The features are extracted from the defect candidate region and the useful combination of features are determined using feature selection. This procedure is applied to each classifier of multiple classifiers. SVM is used to learn the selected features and the discrimination of true defect and pseudo defect is done based on the results of each SVM.

2.1 Definition of Class in Defect of Electronic Board

Iwahori *et al.* [12] classifies into two classes of true and psuedo defects, while this paper considers an approach to classify into more than three classes. The number of classes is assumed to be three. Class 1 condists of true defects which include lack (Fig.1(a)) and connection, class 2 consists of projection (Fig.1(b)) and wear rust which has the similar intensity as that of lead line, while class 3 includes pseudo defects. True defect is not allowed for the market while pseudo defect is still allowed with cleaning before forwarding to the market. While pseudo defects consists of weak rust (Fig.1(c)) which has the low intensity and dust (Fig.1(d)) which has the high intensity. A defect which is similar to both of true defect and pseudo defect is treated as that in the difficult discrimination class. Not only shape but also intensity are used to classify a defect into each of three classes. When only shape is used, more number of classes should be used but the accuracy for discrimination decreases. This is the reason three classes are introduced in the paper.

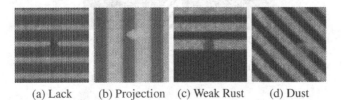

(a) Lack   (b) Projection   (c) Weak Rust   (d) Dust

Fig. 1 Defect of Each Class

2.2 Detection of Defect Candidate Region

Defect candidate is first generated using an inspection image and a reference image respectively. The difference image is generated using an inspection image and a reference image as shown in Fig.4. The reference image is shifted with every one pixel within ± 5 pixels and the minimum point of the shift is determined from the sum of the absolute differences of each image. This process is applied to two reference images. That is, two difference images are generated using two reference images for one inspection image.

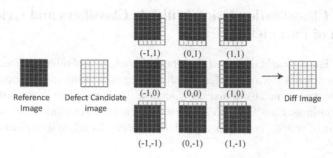

Fig. 2 Obtaining Difference Image

Iwahori *et al.* [12] determines the threshold value empirically for each class of true and pseudo defect, however there is a problem that an appropriate threshold value becomes different based on the condition that the defect is on the lead line or on the base region. The same threshold value also the problem that detected accuracy becomes different. In this situation, this paper proposes a different approach to extract the defect candidate region with higher accuracy.

Discrimination analysis is applied as shown in Fig.3.

(a) Original Image   (b) Binary Image

Fig. 3 Result of Discrimination Analysis

Since the intensity of the board may become different for the same location of the board and the intensity has some range (as shown in Fig.4), the range of intensity taken in the board is used for the threshold value.

Let the maximum intensity of the lead line part be L_h and the minimum intensity be L_l. then the threshold value T_L for the lead line part is represented as Eq.(1).

$$T_L = L_h - L_l \tag{1}$$

Let the maximum intensity of the base part be B_h and the minimum intensity be B_l, then the threshold value T_B is represented as Eq.(2).

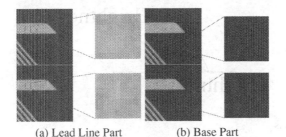

(a) Lead Line Part   (b) Base Part

Fig. 4 Difference of Intensity in Each Part of Board

$$T_B = B_h - B_l \tag{2}$$

Defect candidate region is extractd from pixels which has the difference greater than the threshold T_L in the lead line part and that greater than T_B in the base part. This gives the higher accuracy for extracting the candidate region.

2.3 Feature Extraction and Selection

Features are extracted from the defect candidate region. The features include (1) Mean of Intensity, (2) Maximum Intensity, (3) Minimum Intensity, (4) Proportion of High Intensity, (5) Circularity, (6) Aspect Ratio, (7) Variance, (8) Intensity ratio of candidate region and lead line, (9) Intensity ratio of candidate region and base part, (10) Area, (11) Size of x-direction, (12) Size of y-direction, (13) Perimeter, (14) Diagonal Length, (15) Difference between center of gravity of intensity and maximum intensity, where these features are used in Iwahori *et al.* [12]. Further features used are (16) Complexity, (17) Smoothness, (18) Contrast, (19) Correlation, (20) Angular Second Moment, (21) Inverse Difference Moment, (22) Mode, (23) Skewness, (24) Kurtosis. Thus, a total of 24 features are used. Nine added features of (16) to (24) are used as new features including (18) to (21) in co-occurrence matrix as the statistical features.

2.4 Co-occurrence Matrix

GLCM (Co-occurrence Matrix) uses the relation of relative location of some direction shown in Fig.5 The matrix represents the counts so that the corresponding pair of the pixels (x_1, y_1) and (x_2, y_2) becomes (i, j).

The definitions of features such as contrast, correlation, angular second moment, inverse difference moment obtained from Fig.6 are represented below. $P_x(i)$, $P_y(j)$, μ_x, μ_y, σ_x, σ_y are defined as follows.

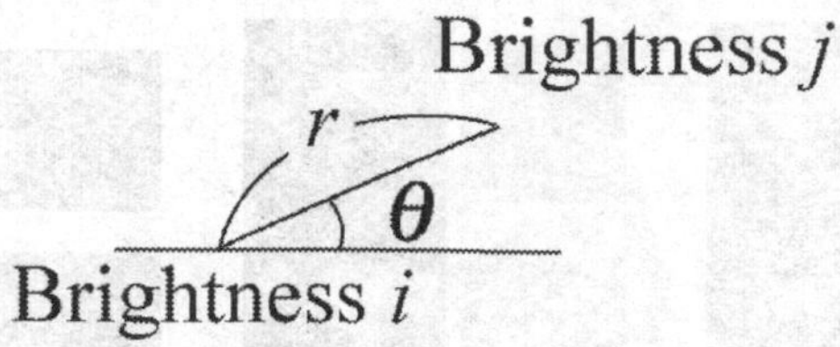

Fig. 5 Pair of Pixels

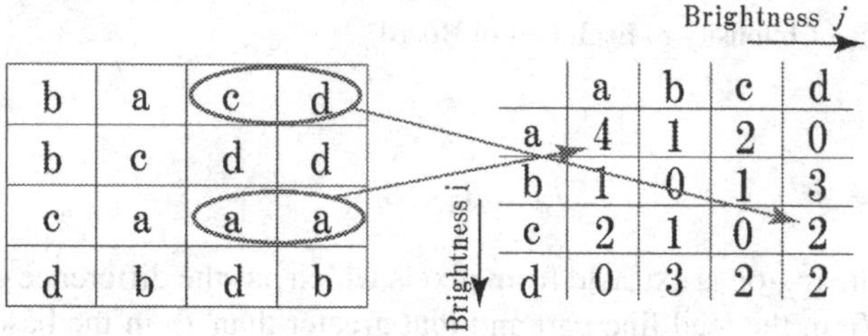

b	a	c	d
b	c	d	d
c	a	a	a
d	b	d	b

	a	b	c	d
a	4	1	2	0
b	1	0	1	3
c	2	1	0	2
d	0	3	2	2

Fig. 6 Co-ocurrence Matrix

$$P_x(i) = \sum_{j=0}^{n-1} P_\delta(i,j) \quad P_y(j) = \Sigma_{i=0}^{n-1} P_\delta(i,j)$$

$$\mu_x = \sum_{i=0}^{n-1} iP_x(i) \quad \mu_y = \Sigma_{j=0}^{n-1} jP_y(j)$$

$$\sigma_x^2 = \sum_{i=0}^{n-1} (i-\mu_x)^2 P_x(i) \ \sigma_y^2 = \Sigma_{j=0}^{n-1} (j-\mu_y)^2 P_y(j)$$

Angular Second Moment (ASM)

$$ASM = \sum_{i=0}^{n-1} \sum_{j=0}^{n-1} (P_\delta(i,j))^2 \tag{3}$$

Contrast (CON)

$$CON = \sum_{i=0}^{n-1} \sum_{j=0}^{n-1} P_\delta(i,j) \tag{4}$$

Correlation (COR)

$$COR = \frac{\Sigma_{i=0}^{n-1} \Sigma_{j=0}^{n-1} ijP_\delta(i,j) - \mu_x \mu_y}{\sigma_x \sigma_y} \tag{5}$$

Inverse Difference Moment (IDM)

$$IDM = \sum_{i=0}^{n-1}\sum_{j=0}^{n-1} \frac{1}{1-(i-j)^2} P_\delta(i,j) \qquad (6)$$

Here, let quantization level be n, and let probability of point which is $\delta = (r, \theta)$ apart from the point of interests with the gray level i be $P_\delta(i,j)$.

2.5 Classification into Multiple Classes

Classification into multiple class determines the class with the distance of hyper plane or at random in general, but this paper treats the defect with difficult discrimination when the final class cannot be determined to one class. This enables to prevent the incorrect classification by keeping the status of difficult discrimination for unknown data which does not belong to one class.

2.5.1 Region of Multiple Classifier

As the representative approaches, there are one-versus-the-rest (1-v-R) and one-versus-one (1-v-1). These approaches cannot determine the final result when the voting number of output becomes the same. This paper proposes a method to improve the problem in Iwahori *et al.* [12]. These two approaches take different region which is used as the difficult classification class. 1-v-1 cannot represent the fuzzy defect for each class. The difference of region is shown in Fig.7.

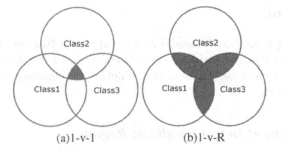

(a)1-v-1   (b)1-v-R

Fig. 7 Region of Same Voting Count for Multiple Classes

2.5.2 Construction of Classifiers

1-v-R used here is shown in Fig.8. 1-v-R constructs M-SVMs $(i = 1,2,...,M)$ to learn the data in M classes and judges the belonged class via integrating its results. Actually there is a case that the same voting count is output to the same class with M classifiers. The approach treats the difficult discrimination to prevent the incorrect classification by keeping the status of difficult discrimination for unknown data which do not belong to one class based on Iwahori *et al.* [12]. In Fig.8, Class 1 represents the true defect which is similar to the intensity of the base part, class 2

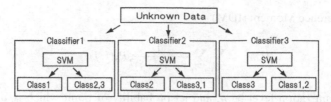

Fig. 8 Construction of Classifier Using 1-v-R

represents the true defect which has the similar intensity to that of the lead line, and class 3 represents the pseudo defect.

2.6   Grid Search

Necessary parameters for SVM are C and σ. Grid search is introduced to determine the combinations of parameters. The method starts the search for wide range of parameters and coarse to fine search is applied to the high precision search with smaller step. Near optimized combination of parameters is determined and the classification can be done with high accuracy as a result.

Here parameter C represents the allowance parameter for incorrect learning. When C takes large value, SVM does not accept the error. Thus the soft margin SVM is introduced with parameter C.

Parameter σ represents the kernel parameter and width of Gaussian kernel. When σ takes large value, region of class becomes wider.

3   Experiment

The proposed approach was compared with the approach Iwahori *et al.* [12]. The approach was also compared with another construction of multiple class classification. The comparison consists of detection of defect candidate region and result of defect classification.

3.1   Extraction of Defect Candidate Region

Result of detection of defect region is shown in (Fig.9 (a)) as an example. The detected image size is 64×64 and the defect candidate region is extracted from the subtraction image (Fig.9 (c)) between an image including defect candidate and a reference image. Discrimination analysis is applied to the reference image to make a binary image (Fig.9 (d)). The threshold values are set for each of lead line part and base part and final defect region (Fig.9 (e)) is extracted from the subtraction image with the determined threshold value. Thus, defect candidate region was extracted with high accuracy (Fig.9 (f)).

Defect candidate region was extracted from a total of four images which consist of two dust images and two weak rust images. Defect candidate region obtained by

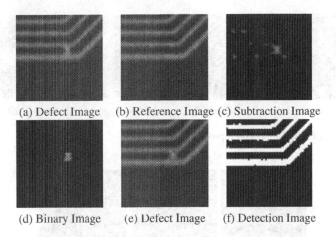

(a) Defect Image   (b) Reference Image (c) Subtraction Image

(d) Binary Image   (e) Defect Image   (f) Detection Image

Fig. 9 Detection Result of [12]

Iwahori *et al.* [12] is shown in Fig.10, while that by the proposed approach is shown in Fig.11. It is observed that Iwahori *et al.* [12] has the cases that the defect candidate region is not always the precise as shown in Fig.10 (a) (b), or over-detection including non-defect region is seen in Fig.10 (c). While it is confirmed that the proposed approach has the higher precision for the detection of defect candidate region.

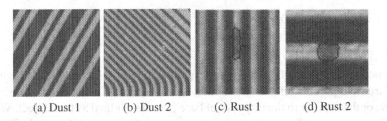

(a) Dust 1   (b) Dust 2   (c) Rust 1   (d) Rust 2

Fig. 10 Extraction by [12]

Other extraction result for other kind of defects except dust and weak rust is shown in Fig.10.

It is shown that other defects are also extracted with high accuracy like dust or weak rust in Fig.12. This result suggests the effectiveness of the automatic determination of threshold value to extract the defect candidate region

3.2 Classification Result

Dataset used consist of 120 images for each of learning data and test data. Features consist of 24 kinds and correct ratio was calculated using feed forward selection for

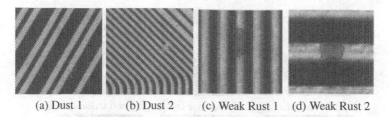

(a) Dust 1   (b) Dust 2   (c) Weak Rust 1   (d) Weak Rust 2

Fig. 11 Extraction by Proposed Approach

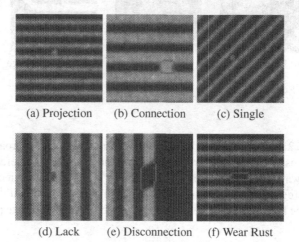

(a) Projection   (b) Connection   (c) Single

(d) Lack   (e) Disconnection   (f) Wear Rust

Fig. 12 Extraction of Each Defect

the learned SVM with RBF kernel. The purpose is to reduce the incorrect classification of true and pseudo defects. When the output is defect similar to the intensity of lead line or that similar to the intensity of base part, it is judged as true defect. While when there are multiple same voting counts, it is judged as a defect with difficult discrimination and the rest is treated as pseudo defect.

Correct ratio is defined as Eq.(7) and correct ratio for positive and negative samples is calculated by Eq.(8) for the evaluation. Further, Iwahori *et al.* [13] considers the incorrect classification for the true defect rather than the incorrect classification for the pseudo defect, and this was evaluated by Eq.(9). Here, let the number of classification of true defect correctly classified be TP, let the number of its incorrect classification be TN, let the number of its difficult classification be TD, let the number of correct classification of pseudo defect be FP, the number of its incorrect classification be FN, the number of its difficult classification be FD.

$$Accuracy = \frac{TP + FP}{TP + TN + FP + FN} \times 100 \quad [\%] \tag{7}$$

$$ErrorateAccuracy = \frac{TP + TN + FP + FN}{AllData} \times 100 \quad [\%] \qquad (8)$$

$$Performance = \frac{AllData - TN}{TP + TN + FP + TD + FD} \times 100 \quad [\%] \qquad (9)$$

The classification accuracy of each classifier is shown in Table I. The result of each classifier in the proposed approach is shown in Table I and the result of classification of unknown data using proposed approach and Iwahori et al. [12] is shown in Table II. The accuracy is calculated from correct classification and incorrect classification except difficult judgment class. represents the portion of data which was not classified to the difficult class. The value of parameter C was 1 and the values of σ were 0.5 for classifier 1, 0.15 for classifier 2, and 0.39 for classifier 3 from the result of grid search.

Table 1 suggests that the efficient feature selection has been done for each classifier based on the different combination of features used in each classifier. Added features selected are contrast(18)Ccorrelation(19), and inverse difference moment(21) and these features are effective to the defect classification of electronic board.

Table 2 gives comparison between the proposed approach and Iwahori et al. [12] shows that 6.6% is improved for the accuracy, 43% is improved for the correct ratio for positive and negative samples, and 5.2% is improved for the performance with

Table 1 Result of Each Classifier

		Classifier1	Classifier2	Classifier3
Feature		7.8.9.19	8.9	8.9.18.21
First Class	True	96	91	88
	False	4	9	12
Second Class	True	195	194	188
	False	5	6	12
Accuracy		97.0%	95.0%	92.0%
Performance		98.6%	96.9%	95.8%

Table 2 Comparison between Proposed Approach and [12]

		Proposed Method	1-v-1 Mehod	Reference[11]
True Defect	True	183	184	89
	False	3	14	9
	Difficult	14	2	102
Pseudo Defect	True	86	85	45
	False	5	15	5
	Difficult	9	0	50
Accuracy		97.1%	89.9%	90.5%
Errorate Accuracy		92.3%	99.3%	49.3%
Performance		98.9%	95.1%	93.7%

the decrease of the incorrect classification. This is based on defining one class from the result of multiple class classification not treating one class for the margin region.

Comparison between the proposed approach and 1-v-1 approach, 1-v-R gave 8 incorrect samples but 1-v-1 gave 29 incorrect samples based on the reason without covering the region over the classes.

The proposed approach introduced a class for a defect with difficult sidcrimination in order to reduce the error rate for the true defects. The effect is confirmed from the evaluation result.

3.3 Incorrect Classification

Incorrect classification of true defect is a problem rather than that of pseudo defect in the defect classification of electronic board. Sample images in the incorrect classification of true defect to pseudo defect is shown in Fig.13.

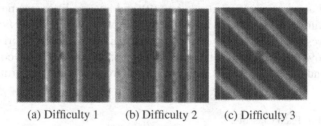

(a) Difficulty 1   (b) Difficulty 2   (c) Difficulty 3

Fig. 13 Incorrect Classification of True Defect

Fig.13(a) or Fig.13(b) are the lack of true defect and its width depends on the shape of the defect. It tends to be misclassified for the defect which has the small area of defect candidate region. The defect obtained from the defect candidate region is not necessarily misclassified and single defect is not misclassified. In case that there are similar data, the classification tends to be succeeded. Fig.13(c) shows the connected defect of true and pseudo defects which have the similar intensity values. This defect was treated as one defect and this caused misclassification.

4 Conclusion

This paper proposed a new approach to improve the classification accuracy of defect of electronic board by introducing classification with multiple classes, reducing the number of misclassified samples with adding features and determining parameters of SVM.

The usefulness of the approach was shown through the comparison with the previous approach with the experiments for true defect, pseudo defect and difficult judgment of classification. Parameters of SVM were determined by discrimination analysis and grid search. The evaluation of the proposed approach provided 97%

accuracy with the dataset. Reducing the number of misclassification and recognition percentage of true defect to the pseudo defect are necessary and this is further subject.

Acknowledgment. Iwahori's research is supported by JSPS Grant-in-Aid for Scientific Research (C) (23500228) and Chubu University Grant.

References

1. Tanaka, T., Hotta, S., Iga, T., Nakamura, T.: Automatic Image Filter Creation System: To Use for a Defect Classification System. IEICE Technical Report 106(448), 195–198 (2007)
2. Rau, H., Wu, C.-H.: Automatic Optical Inspection for Detecting Defects on Printed Circuit Board Inner Layers. The International Journal of Advanced Manufacturing Technology 25(9-10), 940–946 (2005)
3. Kondo, K., et al.: Defect Classification Using Random Feature Selection and Bagging. The Journal of the Institute of Image Electronics Engineers of Japan 38(1), 9–15 (2009) (in Japanese)
4. Onishi, H., Sasa, Y., Nagai, K., Tatsumi, S.: A Pattern Defect Inspection Method by Grayscale Image Comparison without Precise Image Alignment. J. IEICE, D-II J86-D-II(11), 1531–1545 (2003)
5. Maeda, S., Ono, M., Kubota, H., Nakatani, M.: Precise Detection of Short Circuit Defect on TFT Substrate by Infrared Image Matching. J. IEICE D-II J80-D-II(9), 2333–2344 (1997)
6. Numada, M., Koshimizu, H.: A Method for Detecting Globally Distributed Defects by Using Learning with Mahalanobis Distance. In: ViEW 2007, pp. 9–13 (2007)
7. Kondo, K., Kikuchi, K., Hotta, S., Shibuya, H., Maeda, S.: Defect Classification Using Random Feature Selection and Bagging. IIEEJ 38(1), 9–15 (2009)
8. Wakabayashi, T., Tsuruoka, S., Kimura, F., Miyake, Y.: Study on Feature Selection in Handwritten Numeral Recognition. IEICE J78-D-II(11), 1627–1638 (1995)
9. Ishii, K., Ueda, N., Maeda, E., Murase, Y.: Descriptive Pattern Recognition. Ohmsha (1998)
10. Amabe, H., Nagao, T.: A Support Vector Machine Approach to Defect Classification by Selecting Image Characteristics. In: FIT 2006, vol. I-033, pp. 79–80 (2006)
11. Roh, B., Yoon, C., Ryu, Y., Oh, C.: A Neural Network Approach to Defect Classification on Printed Circuit Boards. JSPE 67(10), 1621–1626 (2001)
12. Iwahori, Y., Futamura, K., Adachi, Y.: Discrimination of True Defect and Indefinite Defect with Visual Inspection Using SVM. In: König, A., Dengel, A., Hinkelmann, K., Kise, K., Howlett, R.J., Jain, L.C. (eds.) KES 2011, Part IV. LNCS, vol. 6884, pp. 117–125. Springer, Heidelberg (2011)
13. Iwahori, Y., Kumar, D., Nakagawa, T., Bhuyan, M.K.: Improved Defect Classification of Printed Circuit Board Using SVM. In: KES IDT 2012, pp. 1–10 (2012)

Speeding Up Statistical Tests to Detect Recurring Concept Drifts

Paulo Mauricio Gonçalves Júnior and Roberto Souto Maior de Barros

Abstract. RCD is a framework for dealing with recurring concept drifts. It reuses previously stored classifiers that were trained on examples similar to actual data, through the use of multivariate non-parametric statistical tests. The original proposal performed statistical tests sequentially. This paper improves RCD to perform the statistical tests in parallel by the use of a thread pool and presents how parallelism impacts performance. Results show that using parallel execution can considerably improve the evaluation time when compared to the corresponding sequential execution in environments where many concept drifts occur.

Keywords: Data streams, recurring concept drifts, multivariate non-parametric statistical tests, parallelism.

1 Introduction

Concept drift is a common situation when dealing with data streams. Several authors have defined it in different terms. One of these definitions was stated by Wang et al. [23]: "the term concept refers to the quantity that a learning model is trying to predict, i.e., the variable. Concept drift is the situation in which the statistical properties of the target concept change over time." Kolter and Maloof offered a more informal definition: "concept drift occurs when a set of examples has legitimate class labels at one time and has different legitimate labels at another time" [17].

Paulo Mauricio Gonçalves Júnior
Instituto Federal de Educação, Ciência e Tecnologia de Pernambuco, Cidade Universitária, 50.740-540, Recife, Brasil
e-mail: paulogoncalves@recife.ifpe.edu.br

Roberto Souto Maior de Barros
Centro de Informática, Universidade Federal de Pernambuco, Cidade Universitária, 50.740-560, Recife, Brasil
e-mail: roberto@cin.ufpe.br

R. Lee (Ed.): *Computer and Information Science*, SCI 493, pp. 129–142.
DOI: 10.1007/978-3-319-00804-2_10   © Springer International Publishing Switzerland 2013

Concept drifts may occur in several different situations, in applications such as spam filtering [6], credit card fraud detection [22], and intrusion detection [18].

In recent years, many proposals have been made to deal with concept drifts, like the use of concept drift detectors and ensemble classifiers. One actual solution to deal with recurring concept drifts, named RCD, was previously proposed to perform non-parametric multivariate statistical tests to identify if a concept is recurring or not, and if so, reuse the classifier built on similar data.

In this paper, we present the results of executing the statistical tests in parallel: how much faster it is when compared to sequential execution, in which situations it reports better results, the influence of abrupt and gradual concept drifts in the test results, and how RCD performs in environments with different number of processing cores.

The rest of this paper is organized as follows: Sect. 2 presents some common techniques used to deal with concept drifts; Sect. 3 summarizes the RCD framework and how the parallelism was implemented; Sect. 4 describes the data sets used and their parameters, the evaluation methodology, the RCD configuration, and other information about the experiments; Sect. 5 introduces the results of the experiments; and, finally, Sect. 6 presents our conclusions.

2 Background

There are many approaches used to deal with concept drifts. One approach is to create a single classifier that adapts its internal structure as new data arrive. A commonly used single classifier is based on a Hoeffding tree [7], also named VFDT (Very Fast Decision Tree). It is a decision tree that uses a Hoeffding bound to calculate how much data it needs to process in order to select the value of a decision node. Accuracy of results is similar to that of a batch decision tree, but using much less memory. In its original form, it was not designed to handle concept drifts. Many extensions have already been proposed to adapt Hoeffding trees to deal with concept drifts.

One of these proposals is named CVFDT (Concept-adapting Very Fast Decision Tree) [16]. It states that CVFDT "is an extension to VFDT which maintains VFDT's speed and accuracy advantages but adds the ability to detect and respond to changes in the example-generating process". It uses a sliding window of examples to try to keep its model up-to-date. For each new arriving instance, statistics are recomputed, reducing the influence of older instances. When the concept begins to change, alternative attributes increase their information gain, making the Hoeffding test on the split to fail. An alternative tree begins to grow with the new best attribute at its root. If this subtree becomes more accurate than the old one on new data, it is substituted.

VFDTc [13], on the other hand, extends VFDT with the ability to deal with numeric attributes and uses naive Bayes classifiers at tree leaves. Proposals with decision rules were also made [9].

Another common approach to deal with a concept drift is to identify when it occurs and create a new classifier. Therefore, only classifiers trained on a current concept are maintained. Algorithms that follow this approach work in the following way: each arriving training instance is first evaluated by the base classifier. Internal statistics are updated with the results and two thresholds are computed: a warning level and an error level. As the base classifier makes mistakes, the warning level is reached and instances are stored. If the behavior continues, the error level will be reached, indicating that a concept drift has occurred. At this moment, the base classifier is destroyed and a new base classifier is created and initially trained on the stored instances. On the other hand, if the classifier starts to correctly evaluate instances, this situation is considered a false alarm and stored instances are flushed. Algorithms that follow this approach can work with any type of classifier as they only analyze how the classifier evaluates instances.

One example of this approach is DDM (Drift Detection Method) [10]. It works by controlling the algorithm's error rate. For each point i in the sequence of arriving instances, the error rate is computed as the probability of misclassifying (p_i), with standard deviation given by $s_i = \sqrt{p_i(1 - p_i)/i}$. Statistical theory guarantees that, when the distribution changes, the error will increase. The values of p_i and s_i are stored when $p_i + s_i$ reaches its minimum value during the process (obtaining p_{min} and s_{min}). The warning level is reached when $p_i + s_i \geq p_{min} + 2 \times s_{min}$ and the error level is set at $p_i + s_i \geq p_{min} + 3 \times s_{min}$.

Another similar method is Early Drift Detection Method (EDDM) [1]. It works similarly to DDM, but, instead of controlling solely the amount of error of the classifier, it uses the distance between two errors to identify concept drifts. It computes the average distance between two errors (p_i) and the standard deviation of p_i (s_i). These values are stored when $p_i + 2 \times s_i$ reaches its maximum value (obtaining p_{max} and s_{max}). Thus, the value of $p_{max} + 2 \times s_{max}$ corresponds to the point where the distribution of distances between errors is maximum. EDDM was shown to be more adequate to detect gradual concept drifts while DDM was better suited for abrupt concept drifts [1].

Exponentially weighted moving average (EWMA) charts [19] were originally proposed for detecting an increase in the mean of a sequence of random variables, considering that the mean and standard deviation of the stream are known. Yeh et al. [25] proposed an EWMA change detector for a sequence of random variables that form a Bernoulli distribution. ECDD (EWMA for Concept Drift Detection) [20], extends EWMA to monitor the misclassification rate of a streaming classifier, allowing the rate of false positive detection to be controlled and kept constant over time.

Several proposals try to deal with concept drifts by the use of ensemble classifiers. This approach maintains a collection of learners and combine their decisions to make an overall decision. To deal with concept drifts, ensemble classifiers must take into account the temporal nature of the data stream.

LEARN^{++}.NSE is a recent proposal of an ensemble classifier. The original algorithm [8] works as follows: a single classifier is created for each data set that becomes available. The algorithm first evaluates the classification accuracy of the current ensemble on the newly available data, obtained by the weighted majority voting of all classifiers in the ensemble. Its error is computed as a simple ratio of the correctly identified instances of the new data set and normalized in the interval

[0,1]. Then, the weight of the instances are updated: the weights of the instances misclassified by the ensemble are reduced by a factor of the normalized error. The weights are then normalized; a new classifier is created; and all the classifiers generated so far are evaluated on the current data set, by computing their weighted error. If the error of the most recent classifier is greater than 0.5, it is discarded and a new one is created. For each of the other classifiers, if its error is greater than 0.5, its voting power is removed during the weighted majority voting.

Another proposal for ensemble classifier is DWAA (Dynamic Weight Assignment and Adjustment) [24]. It creates classifiers based on data chunks, using the next chunk to evaluate the classifier previously built. If the ensemble is not full, the classifier is added; otherwise, the worst classifier in the last data chunk is replaced. To set the weight, it uses a formula that considers how many of the ensemble classifiers have actually made correct predictions. If more than half of the classifiers predictions are correct, each one receives a normal reward. Otherwise, each one receives a higher reward, making those influence more the global decision of the ensemble, as they are better suited to represent the concept.

3 Parallel RCD

RCD [14, 15] is a framework developed to deal with recurring concept drifts. It keeps a collection of pairs of classifiers and samples used to train these classifiers, as presented in Fig. 1. In the training phase, a concept drift detector is used. If it identifies a concept drift, a multivariate non-parametric statistical test is performed to compare actual data to stored samples. If the statistical test informs that both data come from the same distribution, the classifier associated with the stored sample is reused, meaning that the classifier is adequate to deal with actual data.

On the other hand, if the test indicates that samples are not similar, the next stored data sample is used for testing, and so on. If no stored classifier is apt for actual data, a new classifier is created and stored in the set. If the set is full, the older classifier is substituted. In the testing phase, statistical tests are performed every t instances (a user parameterized value) to select, from the stored classifiers, the best one for actual data. Thus, RCD dynamically adapts to the current data distribution even in the testing phase.

Originally, RCD performed the statistical tests sequentially. Thus, a statistical test would be performed comparing actual data to data stored in the buffer for classifier 1 to verify if both represented the same data distribution. If positive, this classifier

Fig. 1 RCD classifiers set

was considered the new actual classifier; else, a statistical test would be performed on classifier 2 buffer data, and so on.

The improvement being proposed is to perform several tests *simultaneously* by the use of a thread pool of configurable fixed size to allow the user to fine tune its value based on the hardware being used. Fig. 2 presents an example illustrating how the thread pool works. It considers a thread pool with two active cores and a classifiers set of size six.

When a concept drift occurs, it means the actual classifier does not correctly represent the actual context. So, it is necessary to check whether any stored classifier better represents the actual context. The remaining five classifiers stored in the set must be tested, comparing a sample of actual data to the data stored in the buffer associated with each classifier which represents the data the classifier was trained on. Five threads are built to perform the statistical tests and they are sent to the thread pool using a FIFO scheme to associate each test to a position in the thread pool, but only the first two are active, i.e., are actually performing a statistical test. At Fig. 2 they are represented by bolder lines, and inactive threads by thinner lines. At this point ($t = 0$), two statistical tests are active and the remaining three are waiting to execute.

When the first statistical test finishes (let's consider statistical test 1), if the result indicates that actual data and sample data from classifier 1 do not represent the same data distribution, the next inactive statistical test (in this case, statistical test 3) executes in the corresponding slot ($t = 1$). At $t = 2$, the same occurs. Classifier 2, represented by statistical test 2, also does not better represent actual data and the next statistical test (number 4) occupies its place.

Now, let's consider that statistical test 3 has finished and it identified that actual data and data stored in the buffer of classifier 3 represent the same distribution. In

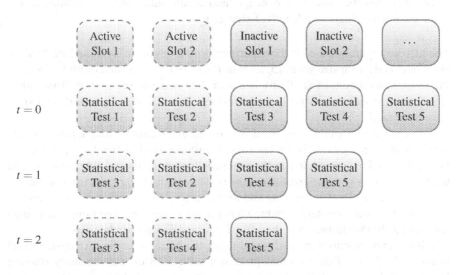

Fig. 2 Example of a thread pool execution

this situation, this classifier substitutes the actual classifier, all other active statistical tests are stopped from executing, and the inactive ones are canceled.

This scheme is interesting because, if a test is negative, the next test to perform is already being executed, allowing a faster performance of the algorithm. If a test is positive, all other executing tests are stopped and tests yet to be executed do not enter the active thread pool.

Notice that this scheme is general and allows the execution of any statistical test in parallel. Source code and instructions on how to use RCD are available as a MOA extension and can be obtained at http://sites.google.com/site/moaextensions/.

4 Experiments Configuration

We used several data sets to perform the experiments: Hyperplane [16], LED [11], SEA [21], Forest Covertype [4], Poker Hand [3], and Electricity [10, 12]. The first three are artificial data sets: the first one presents gradual concept drifts while the following two present abrupt concept drifts. The last three are real-world data sets.

These data sets and their configurations are the same as used by Bifet et al. [3]. Hyperplane was tested in ten million instances while LED and SEA, in one million. All tests in the artificial data sets were repeated ten times and computed a 95% confidence interval. The parameters of these streams are the following:

- HYP(x, v) represents a Hyperplane data stream with x attributes changing at speed v;
- LED(v) appends four concepts $(1, 3, 5, 7)$, each one representing a different number of drifting attributes with length of change v;
- SEA(v) uses the same four concepts and in the same order as defined in the original paper [21], with length of change v.

The RCD configuration used in the experiments includes naive Bayes as base learner, classifiers collection size set to 15, KNN as the statistical test used (with $k = 3$), and the minimum amount of similarity between data samples set to 0.05. Two buffer sizes, two test frequencies (only in the testing phase), and three thread pool sizes have been used.

The evaluation methodology used was Interleaved Chunks, also known as data block evaluation method [5], on ten runs. It initially reads a block of d instances. When the block is formed, it uses the instances for testing the existing classifier and then the classifier is trained on the instances. This methodology was used because it is better suited to compute training and testing times. In the following experiments, d was set to 100,000 instances in the Hyperplane, Covertype, and Poker Hand data sets, and to 10,000 instances in the LED, SEA, and Electricity data sets.

All the experiments were performed using the Massive Online Analysis (MOA) framework [2] in a Core i3 330M processor with 4GB of main memory running Windows 7 Professional. This processor has four cores, two physical and two virtual ones, where each core runs at 2.13GHz.

We used a modified version of the Interleaved Chunks presented in the MOA framework because the version available in the tool computes the time the thread executing the classifier uses the processor. In this solution, it is possible to execute other applications at the same time and the results will not be affected. However, because we will use a thread pool to perform the statistical tests, these are not computed because the original thread may not be active in the processor. Here, we compute the real time taken by RCD to perform, not being possible to run other applications at the same time.

5 Results

Table 1 presents the average number of detected concept drifts, classifiers set size, number of reused classifiers, as well as the evaluation, train and test times (in seconds) for RCD considering the ten runs, using a buffer size with 100 instances and a test frequency of 500, considering thread pools with one, two, and four active cores.

It is worth pointing out the results were quite similar in the artificial data sets, regardless of the thread pool size. This behavior did not occur in the real-world data sets and is probably related to the number of detected concept drifts. In the artificial data sets, the average number of detected concept drifts are considerably low, as can be seen in the first column (CD), because a small number of concept drifts demand few statistical tests to be performed.

However, not only the number of detected concept drifts influences performance. Reusing classifiers also matters. If the first tests identify similarity between distributions, several other tests will not be executed, reducing the benefits of using a thread pool. On the other hand, if only the last tests or none at all identify similarity, more tests need to be executed. This is the situation expected to be more benefited from the parallelization of the statistical tests.

For example, the average classifiers collection size (CS) in the artificial data sets is below three, not being even close to fill the set (15 classifiers). Having few stored classifiers indicates that few statistical tests need to be performed. The difference between the number of detected concept drifts and the number of stored classifiers is due to the reuse of classifiers. Analyzing the column with the number of reused classifiers (RC), we can see that the values are very close to the ones presented in the first column. This means that in the majority of the concept drifts a classifier was reused.

In this RCD configuration, analyzing the artificial data sets, using one core had better statistical results than using two cores in both configurations of Hyperplane and in LED but worse results in both versions of SEA. Using one core performed statistically better in the HYP(10, 0.0001) and LED data sets when compared to using four cores but worse performance in SEA, similarly to using two cores. In the HYP(10, 0.001) data set, both had statistically similar results. Using four cores had better statistical results than using two cores in the Hyperplane and SEA data sets, and similar ones in LED.

Table 1 Results for a buffer with 100 instances and test frequency of 500 instances (in seconds)

Data sets	CD	CS	RC	1 core			2 cores			4 cores		
				eval	train	test	eval	train	test	eval	train	test
HYP(10,0.001)	7.7	2.6	6.0	99.64	44.45	36.30	101.49	45.40	39.98	100.13	46.13	38.20
HYP(10,0.0001)	8.2	2.4	6.7	98.99	44.06	36.15	101.49	45.35	40.29	100.10	46.18	38.18
SEA(50)	0.1	1.1	0.0	4.70	2.11	1.09	4.32	1.53	1.31	4.30	1.56	1.27
SEA(50000)	0.4	1.1	0.3	4.63	1.97	1.25	4.26	1.54	1.31	4.23	1.56	1.27
LED(50000)	0.3	1.2	0.1	32.56	14.23	12.78	33.54	14.72	13.25	33.54	14.70	13.24
Covertype	2980.0	15.0	844.0	98.12	80.68	8.30	84.24	66.99	8.25	73.99	56.77	8.32
Poker Hand	1871.0	15.0	109.0	45.72	37.86	3.78	38.31	30.45	3.81	31.67	23.82	3.82
Electricity	212.0	15.0	30.0	2.89	2.48	0.09	2.40	1.98	0.09	2.00	1.58	0.09

Table 2 Thread pool management for a buffer with 100 instances and test frequency of 500 instances (in milliseconds)

	Creation time			Execution time			Destruction time		
	1 core	2 cores	4 cores	1 core	2 cores	4 cores	1 core	2 cores	4 cores
HYP(10,0.001)	2.49	1.03	1.03	1.87	2.22	2.63	0.00	0.00	0.00
HYP(10,0.0001)	2.49	1.16	0.40	1.73	2.90	3.68	0.00	0.00	0.00
SEA(50)	0.00	0.00	0.00	0.00	0.00	0.00	0.00	0.00	0.00
SEA(50000)	4.00	0.00	0.00	0.00	0.00	4.00	0.00	0.00	0.00
LED(50000)	0.00	5.33	0.00	5.00	0.00	10.33	0.00	0.00	0.00
Covertype	2.12	2.25	2.97	25.76	19.84	15.04	0.03	0.02	0.04
Poker Hand	2.01	2.21	2.96	14.81	10.89	6.10	0.01	0.01	0.06
Electricity	2.16	1.95	2.92	8.82	6.58	3.42	0.00	0.00	0.00

Table 3 Results for a buffer and test frequency of 100 instances (in seconds)

Data sets	CD	CS	RC	1 core			2 cores			4 cores		
				eval	train	test	eval	train	test	eval	train	test
HYP(10,0.001)	8.4	2.8	6.5	422.74	45.61	360.38	487.43	46.86	424.62	499.94	46.87	437.12
HYP(10,0.0001)	10.2	2.3	8.2	421.24	45.37	359.88	459.86	46.66	397.05	454.82	46.63	392.47
SEA(50)	0.1	1.1	0.0	6.18	1.51	3.15	6.25	1.55	3.19	6.22	1.54	3.16
SEA(50000)	0.2	1.1	0.1	6.27	1.53	3.23	6.35	1.56	3.29	6.34	1.58	3.25
LED(50000)	2.3	1.2	2.1	37.22	14.25	17.38	38.47	14.89	17.98	38.51	14.90	18.06
Covertype	3376.0	15.0	944.0	534.32	86.47	438.71	354.76	72.31	273.38	315.25	60.87	245.36
Poker Hand	1871.0	15.0	109.0	305.93	37.83	264.08	229.43	30.78	194.59	166.22	23.65	138.53
Electricity	212.0	15.0	30.0	12.40	2.43	9.62	9.50	2.07	7.08	6.80	1.50	4.96

Table 4 Thread pool management for a buffer and test frequency of 100 instances (in milliseconds)

	Creation time			Execution time			Destruction time		
	1 core	2 cores	4 cores	1 core	2 cores	4 cores	1 core	2 cores	4 cores
HYP(10,0.001)	0.39	0.49	0.51	2.85	3.36	3.47	0.01	0.01	0.01
HYP(10,0.0001)	0.32	0.37	0.36	2.91	3.21	3.17	0.01	0.01	0.01
SEA(50)	0.03	0.03	0.02	0.15	0.15	0.15	0.00	0.00	0.00
SEA(50000)	0.02	0.03	0.03	0.16	0.16	0.16	0.00	0.00	0.00
LED(50000)	0.03	0.03	0.04	0.40	0.41	0.41	0.00	0.00	0.00
Covertype	2.05	2.35	3.14	73.84	46.14	39.73	0.01	0.02	0.07
Poker Hand	2.14	2.39	3.08	30.97	21.95	14.00	0.01	0.02	0.08
Electricity	2.08	2.75	2.85	21.79	15.07	9.69	0.08	0.00	0.03

On the other hand, in the real-world data sets, more cores returned lower evaluation times: two cores were faster than one core and using four cores was faster than the other two thread pool sizes. In the artificial data sets, using one core was, on average, 1.90% faster than using two cores, but 14.84% slower in the real-world data sets. Comparing one and four cores, similar results apply. One core was faster by 0.74% in the artificial data sets while four cores was faster by 26.63% in the real data sets. Using four cores had practically the same performance than using two cores in artificial data sets, being 1.14% faster, but was 13.85% faster in the real-world data sets. The real-world data sets presented a huge amount of concept drifts, the classifiers set became full and the number of reused classifiers was also much bigger than in the artificial data sets.

To better analyze how the thread pool influences performance, we computed the average amount of time (in milliseconds) needed to create the thread pool and to assign the statistical tests to their respective slots, to execute the thread pool, and to finalize it. In the assignment stage, if a statistical test is assigned to an active slot, it starts executing immediately, while other tests are still being assigned, so it is not necessary to wait for all tests to be assigned a specific slot to start execution, saving time. This information is presented at Table 2.

Observing the real-world data sets, it is possible to notice that, in general, the greater the number of cores, the longer was the time spent in the creation of the thread pool, but the differences are usually very small. In the execution time, using more cores meant faster execution, with no exception. The destruction times were usually negligible, taking less than 0.1 milliseconds.

The results of Table 3 are similar to those presented at Table 1 but the test frequency in the experiments was increased to 100 instances. Again, parallelism outperforms sequential solution in the real-world data sets, the ones with more detected concept drifts.

In these tests, we can also notice that the evaluation time is mostly spent in the testing phase, differently from the results of Table 1. As the test frequency is higher, the evaluation time and the time spent in the testing phase increased considerably. Making the tests more frequently also increased the number of detected concept drifts in 50% of the data sets. In SEA(50,000), it was 0.4 to 0.2. In Poker Hand, Electricity, and SEA(50), the number of detected concept drifts stayed the same.

Using one core was faster than using two cores in average by 11.72% in the artificial data sets, but was 30.37% slower in the real-world data sets. Comparing one core to four cores similar results apply: one core was faster by 12.55% in the artificial data sets and four cores was faster by 42.74% in the real-world data sets. Using two cores offered better average results compared to using four cores by 0.75% in the artificial data sets and worse performance by 17.76% in the real-world ones.

Table 4 presents similar information to those presented at Table 2. Results are also similar: the creation times is usually slightly faster using less cores, the execution time is considerably smaller when using more cores, and destruction times are usually less than 0.1 milliseconds.

Instead of increasing the test frequency, Table 5 presents similar information as Tables 1 and 3 but increasing the buffer size to 500 instances. Here, the tests took

Table 5 Results for a buffer and test frequency of 500 instances (in seconds)

Data sets	CD	CS	RC	1 core			2 cores			4 cores		
				eval	train	test	eval	train	test	eval	train	test
HYP(10,0.001)	11.9	2.6	9.9	1339.33	46.07	1275.16	1453.37	46.65	1388.60	1508.29	48.98	1441.27
HYP(10,0.0001)	9.5	2.4	7.5	1356.57	45.77	1293.32	1404.87	46.01	1341.41	1413.77	47.41	1348.88
SEA(50)	1.1	1.1	1.0	11.17	1.59	7.30	11.17	1.57	7.31	11.25	1.62	7.30
SEA(50000)	0.2	1.1	0.1	11.39	1.56	7.57	11.40	1.58	7.55	11.49	1.60	7.60
LED(50000)	0.2	1.2	0.0	57.99	14.28	37.33	53.98	14.25	33.33	55.14	14.92	33.85
Covertype	3063.0	15.0	798.0	1781.17	358.50	1413.39	1107.10	244.95	853.07	1167.37	268.35	889.92
Poker Hand	410.0	15.0	18.0	588.42	47.24	537.02	359.69	34.10	321.47	377.63	36.19	336.09
Electricity	183.0	15.0	20.0	33.38	8.85	24.15	20.87	6.02	14.46	22.92	7.19	15.26

Table 6 Thread pool management for a buffer and test frequency of 500 instances (in milliseconds)

	Creation time			Execution time			Destruction time		
	1 core	2 cores	4 cores	1 core	2 cores	4 cores	1 core	2 cores	4 cores
HYP(10,0.001)	0.54	0.60	0.60	61.90	67.54	70.14	0.02	0.02	0.02
HYP(10,0.0001)	0.50	0.55	0.51	62.94	65.31	65.66	0.02	0.01	0.02
SEA(50)	0.04	0.03	0.04	2.99	2.99	2.98	0.00	0.00	0.00
SEA(50000)	0.04	0.05	0.04	3.10	3.09	3.09	0.00	0.00	0.01
LED(50000)	0.06	0.04	0.05	12.32	10.27	10.26	0.00	0.00	0.00
Covertype	2.01	2.34	13.66	560.81	342.49	354.36	0.02	0.01	0.06
Poker Hand	2.27	2.65	6.36	314.13	186.57	187.92	0.01	0.06	0.03
Electricity	2.04	2.28	5.23	140.45	91.20	91.78	0.00	0.06	0.06

longer to complete; in average, 62 milliseconds compared to 3 milliseconds when using a buffer with 100 instances. Nevertheless, the results were similar to the ones presented at Table 3. Using parallelism was much faster in the real-world data sets and slightly slower in the artificial ones. Using one core was 5.70% faster than using two cores in the artificial data sets but 38.09% slower in the real-world ones.

However, it was interesting to observe that the evaluation time was lower using two cores than with four. Comparing one and four cores similar results apply: one core was 8.05% faster in the artificial data sets and 34.75% slower in the real-world ones. Using two cores was slightly better than using four: 2.22% in the artificial and 5.39% in the real-world data sets. This probably occurs because there are much more statistical tests to perform and they take longer to complete than the tests in the other configurations, putting a higher load on the whole system and negatively affecting the performance.

Comparing Tables 1, 3 and 5, it is possible to notice that the increase in the buffer size had a higher influence in the evaluation time than the increase in the test frequency. Increasing the test frequency by five times increased the evaluation time between 4.26 and 4.50 times. Increasing the buffer size by five times increased the evaluation time between 11.95 and 13.37 times. The training time practically did not change in the three configurations performed; the increase in the evaluation time was due to the testing time.

Table 6 presents the times taken for the thread pool management, as previously described at Tables 2 and 4. In the artificial data sets, the creation times are very close in the three sizes of active cores used. In the real-world data sets, one and two cores take very similar times, and using four cores takes more time than the other two. This probably occurs because, in the creation time, the statistical tests associated with active cores begin executing while other tests are still being assigned. Thus, the creation time tends to be bigger when there are more active cores and they take longer to complete. We can see it comparing the three tables concerning thread pool management. At Tables 2 and 4, the differences in the creation time between one, two and four cores are almost negligible. In these cases, the average time taken to perform a statistical test is three milliseconds. At Table 6, using four cores takes more time than using one and two cores. Here, the average time taken to perform the statistical test is 62 milliseconds. The execution times are very similar in the artificial data sets. In the real-world data sets, using two or four cores is considerably faster than using one active core. Using two cores was faster than using four cores in the Covertype data set, while in the other two real-world data sets, the performances were quite similar.

6 Conclusion

This paper studied the influence of executing parallel statistical tests in the RCD framework using six data sets (eight configurations), with and without concept drifts, with abrupt and gradual concept drifts, and considering artificial and real-world data sets. Tests were performed with both sequential and parallel execution of two and four statistical tests.

Analysis of the experiment results led to the conclusion that the execution of parallel statistical tests was most beneficial when there was a high number of detected concept drifts leading to more statistical tests being performed. Tests were also performed to analyze the performance results in the following conditions:

1. the buffer size was increased, making the statistical tests take longer to complete; and
2. increase the test frequency, making more statistical tests to be performed.

In data sets with a small number of detected concept drifts (the artificial data sets), the performances were quite similar, but using sequential execution had lower evaluation times ranging from 0.74% to 12.55%. On the other hand, in the data sets with a high number of detected concept drifts (the real-world ones), using parallelism increased performance in values ranging from 13.85% to 42.74%.

Multiplying the test frequency five times increased the evaluation time more than four times, while multiplying the size of the buffer five times increased the evaluation time more than 11 times, indicating that the buffer size has a higher impact in performance than test frequency.

The analysis of the thread pool creation, execution, and destruction times were also performed, showing that, as expected, the major improvement occurs in the execution phase. The creation times using different number of cores are close to one another and the destruction times are commonly negligible, taking less than 0.1 milliseconds.

6.1 Future Work

Some other experiments might be made to better understand how the execution of parallel statistical tests can improve the performance of the RCD framework. One of these experiments is testing the influence of the number of available cores in the processor in performance. Other possible experiment is to analyze the influence of other buffer sizes and test frequencies.

References

1. Baena-García, M., Del Campo-Ávila, J., Fidalgo, R., Bifet, A., Gavaldà, R., Morales-Bueno, R.: Early drift detection method. In: International Workshop on Knowledge Discovery from Data Streams, IWKDDS 2006, pp. 77–86 (2006),
   http://eprints.pascal-network.org/archive/00002509/
2. Bifet, A., Holmes, G., Kirkby, R., Pfahringer, B.: MOA: Massive online analysis. J. of Mach. Learn. Res. 11, 1601–1604 (2010),
   http://portal.acm.org/citation.cfm?id=1859890.1859903
3. Bifet, A., Holmes, G., Pfahringer, B., Frank, E.: Fast perceptron decision tree learning from evolving data streams. In: Zaki, M.J., Yu, J.X., Ravindran, B., Pudi, V. (eds.) PAKDD 2010, Part II. LNCS (LNAI), vol. 6119, pp. 299–310. Springer, Heidelberg (2010), http://dx.doi.org/10.1007/978-3-642-13672-6_30
4. Blackard, J.A., Dean, D.J.: Comparative accuracies of artificial neural networks and discriminant analysis in predicting forest cover types from cartographic variables. Comput. and Electron. in Agric. 24(3), 131–151 (1999),
   http://dx.doi.org/10.1016/S0168-1699(99)00046-0
5. Brzeziński, D., Stefanowski, J.: Accuracy updated ensemble for data streams with concept drift. In: Corchado, E., Kurzyński, M., Woźniak, M. (eds.) HAIS 2011, Part II. LNCS, vol. 6679, pp. 155–163. Springer, Heidelberg (2011),
   http://dx.doi.org/10.1007/978-3-642-21222-2_19
6. Delany, S.J., Cunningham, P., Tsymbal, A., Coyle, L.: A case-based technique for tracking concept drift in spam filtering. Knowl.-Based Syst. 18(4-5), 187–195 (2005), http://dx.doi.org/10.1016/j.knosys.2004.10.002; AI-2004, Cambridge, England, December 13-15 (2004)
7. Domingos, P., Hulten, G.: Mining high-speed data streams. In: Proceedings of the Sixth ACM SIGKDD International Conference on Knowledge Discovery and Data Mining, KDD 2000, New York, NY, USA, pp. 71–80 (2000),
   http://dx.doi.org/10.1145/347090.347107
8. Elwell, R., Polikar, R.: Incremental learning of concept drift in nonstationary environments. IEEE Trans. on Neural Netw. 22(10), 1517–1531 (2011),
   http://dx.doi.org/10.1109/TNN.2011.2160459
9. Ferrer-Troyano, F., Aguilar-Ruiz, J.S., Riquelme, J.C.: Discovering decision rules from numerical data streams. In: Proceedings of the 2004 ACM Symposium on Applied Computing, SAC 2004, New York, NY, USA, pp. 649–653 (2004),
   http://dx.doi.org/10.1145/967900.968036

10. Gama, J., Medas, P., Castillo, G., Rodrigues, P.: Learning with drift detection. In: Bazzan, A.L.C., Labidi, S. (eds.) SBIA 2004. LNCS (LNAI), vol. 3171, pp. 286–295. Springer, Heidelberg (2004),
    http://dx.doi.org/10.1007/978-3-540-28645-5_29
11. Gama, J., Medas, P., Rocha, R.: Forest trees for on-line data. In: Proceedings of the 2004 ACM Symposium on Applied Computing, SAC 2004, New York, NY, USA, pp. 632–636 (2004),
    http://dx.doi.org/10.1145/967900.968033
12. Gama, J., Medas, P., Rodrigues, P.: Learning decision trees from dynamic data streams. In: Proceedings of the 2005 ACM Symposium on Applied Computing, SAC 2005, New York, NY, USA, pp. 573–577 (2005),
    http://dx.doi.org/10.1145/1066677.1066809
13. Gama, J., Rocha, R., Medas, P.: Accurate decision trees for mining high-speed data streams. In: Proceedings of the Ninth ACM SIGKDD International Conference on Knowledge Discovery and Data Mining, KDD 2003, New York, NY, USA, pp. 523–528 (2003), http://dx.doi.org/10.1145/956750.956813
14. Gonçalves Jr., P.M., Barros, R.S.M.: A comparison on how statistical tests deal with concept drifts. In: Arabnia, H.R., et al. (eds.) Proceedings of the 2012 International Conference on Artificial Intelligence, ICAI 2012, vol. 2, pp. 832–838. CSREA Press, Las Vegas (2012)
15. Gonçalves Jr., P.M., Barros, R.S.M.: RCD: A recurring concept drift framework. Pattern Recognit. Lett. (to appear, 2013),
    http://dx.doi.org/10.1016/j.patrec.2013.02.005
16. Hulten, G., Spencer, L., Domingos, P.: Mining time-changing data streams. In: Proceedings of the Seventh ACM SIGKDD International Conference on Knowledge Discovery and Data Mining, KDD 2001, New York, NY, USA, pp. 97–106 (2001),
    http://dx.doi.org/10.1145/502512.502529
17. Kolter, J.Z., Maloof, M.A.: Dynamic weighted majority: An ensemble method for drifting concepts. J. of Mach. Learn. Res. 8, 2755–2790 (2007),
    http://dl.acm.org/citation.cfm?id=1314498.1390333
18. Lane, T., Brodley, C.E.: Approaches to online learning and concept drift for user identification in computer security. In: Agrawal, R., Stolorz, P. (eds.) Proceedings of the Fourth International Conference on Knowledge Discovery and Data Mining, KDD 1998, pp. 259–263. AAAI Press, Menlo Park (1998),
    http://www.aaai.org/Papers/KDD/1998/KDD98-045.pdf
19. Roberts, S.W.: Control chart tests based on geometric moving averages. Technometrics 1(3), 239–250 (1959), http://www.jstor.org/stable/1266443
20. Ross, G.J., Adams, N.M., Tasoulis, D.K., Hand, D.J.: Exponentially weighted moving average charts for detecting concept drift. Pattern Recognit. Lett. 33(2), 191–198 (2012),
    http://dx.doi.org/10.1016/j.patrec.2011.08.019
21. Street, W.N., Kim, Y.: A streaming ensemble algorithm (SEA) for large-scale classification. In: Proceedings of the Seventh ACM SIGKDD International Conference on Knowledge Discovery and Data Mining, KDD 2001, New York, NY, USA, pp. 377–382 (2001),
    http://dx.doi.org/10.1145/502512.502568
22. Wang, H., Fan, W., Yu, P.S., Han, J.: Mining concept-drifting data streams using ensemble classifiers. In: Proceedings of the Ninth ACM SIGKDD International Conference on Knowledge Discovery and Data Mining, KDD 2003, New York, NY, USA, pp. 226–235 (2003), http://dx.doi.org/10.1145/956750.956778

23. Wang, S., Schlobach, S., Klein, M.: Concept drift and how to identify it. Web Semant.:
    Sci., Serv. and Agents on the World Wide Web 9(3), 247–265 (2011),
    http://dx.doi.org/10.1016/j.websem.2011.05.003
24. Wu, D., Wang, K., He, T., Ren, J.: A dynamic weighted ensemble to cope with concept
    drifting classification. In: The 9th International Conference for Young Computer Scien-
    tists, ICYCS 2008, pp. 1854–1859 (2008),
    http://dx.doi.org/10.1109/ICYCS.2008.491
25. Yeh, A.B., Mcgrath, R.N., Sembower, M.A., Shen, Q.: Ewma control charts for monitor-
    ing high-yield processes based on non-transformed observations. International Journal
    of Production Research 46(20), 5679–5699 (2008),
    http://dx.doi.org/10.1080/00207540601182252

A Concurrent Tuple Set Architecture
for Call Level Interfaces

Óscar Mortágua Pereira, Rui L. Aguiar, and Maribel Yasmina Santos

Abstract. Call Level Interfaces (CLI) are low level API aimed at providing services to connect two main components in database applications: client applications and relational databases. Among their functionalities, the ability to manage data retrieved from databases is emphasized. The retrieved data is kept in local memory structures that may be permanently connected to the host database. Client applications, beyond the ability to read their contents, may also execute *Insert, Update* and *Delete* actions over the local memory structures, following specific protocols. These protocols are row (tuple) oriented and, while being executed, cannot be preempted to start another protocol. This restriction leads to several difficulties when applications need to deal with several tuples at a time. The most paradigmatic case is the impossibility to cope with concurrent environments where several threads need to access to the same local memory structure instance, each one pointing to a different tuple and executing its particular protocol. To overcome the aforementioned fragility, a Concurrent Tuple Set Architecture (CTSA) is proposed to manage local memory structures. A performance assessment of a Java component based on JDBC (CLI) is also carried out and compared with a common concurrent approach. The main outcome of this research is the evidence that in concurrent environments, components relying on the CTSA may significantly improve the overall performance when compared with solutions based on standard JDBC API.

Keywords: Call Level Interfaces, concurrency, databases, software architecture.

Óscar Mortágua Pereira · Rui L. Aguiar
Instituto de Telecomunicações, DETI-University of Aveiro, 3810-193 Aveiro, Portugal
e-mail: omp@ua.pt, ruilaa@ua.pt

Maribel Yasmina Santos
Centro Algoritmi, DSI-Univerity of Minho, 4800-058 Guimarães, Portugal
e-mail: maribel@dsi.uminho.pt

R. Lee (Ed.): *Computer and Information Science*, SCI 493, pp. 143–158.
DOI: 10.1007/978-3-319-00804-2_11　　© Springer International Publishing Switzerland 2013

1   Introduction

Database applications comprise at least two main components: database components and application components. In our context, application components are
developed in the object-oriented paradigm and database components rely on the
relational paradigm. The two paradigms are simply too different to bridge seam-
lessly, leading to difficulties informally known as impedance mismatch [1]. The
diverse foundations of both paradigms are a major hindrance for their integration,
being an open challenge for more than 50 years [2]. In order to overcome the impedance mismatch issue, several solutions have emerged such as, embedded SQL
(SQLJ [3]), language extensions (LINQ [4]), Call Level Interfaces [5] (CLI)
(JDBC [6], ODBC [7]), object/relational mappings (O/RM) (Hibernate [8], Top-
Link [9], LINQ) and persistent frameworks (JDO [10], JPA [11], SDO [12],
ADO.NET [13]). Despite their individual advantages, these solutions have not
been designed to manage concurrency on the client side of database applications.
Currently, concurrency is managed by database management systems through
database transactions. Moreover, whenever the same data is needed by different
client-threads, each thread behaves as an independent entity requesting its own
data set. In other words, instead of sharing the data returned by a unique execution
of a Select expression, each thread executes a Select expression independently
from other threads. This leads to a waste of resources, namely it requires more
memory, it requires more power computation, and performance is very probably
affected negatively. Current tools use local memory structures (LMS) to manage
the data returned by Select expressions. Beyond services to read the data kept by
LMS, LMS provide services to execute three additional main protocols on their in-
memory data: update data, insert new data and delete data. Thus, client-
applications are able to update data, insert data and delete data without the need to
explicitly execute Update, Insert and Delete expressions, respectively. Once again,
these protocols are not thread-safe not promoting this way the use of LMS on
concurrent environments. Table 1 presents a typical case where one table attribute
needs to be updated. The value to be used to update the attribute is dependent on
the table primary key (PKs). The left-side column presents the current approach
and the right-side column presents an approach based on thread-safe LMS. When
using the current approach, each thread is created and then it runs (doIt) to execute
a task. Each thread has its own LMS, this way preventing any concurrency at the
LMS level. When using the thread-safe approach, all threads share the same LMS
and update the attribute concurrently. In order to overcome the limitations of CLI,
this paper proposes a Concurrent Tuple Set Architecture (CTSA). The CTSA,
unlike current solutions, provides thread-safe protocols to interact with the data
returned by Select expressions, as shown on right column-side of Table 1.

JDBC and ODBC are two of the most representative standards of CLI. JDBC
and ODBC provide, respectively, ResultSet [14] interface and RecordSet [15]
interface as their internal implementations of LMS.

Table 1 Current and thread-safe approaches

Current Approach	Thread-safe LMS

void begin() {
    foreach thread
      creat thread
        thread.doIt(PKs)
    end
}
void doIt(PKs) {
    LMS=execute Select expression
    while more rows on the LMS
      if PK is in PKs
        then update row
        move to the next row
    end while
 	
}
	void begin() {
	    LMS=execute Select expression
	    foreach thread
	      create thread
	        doIt(LMS,PKs)
	    end
	}
	void doIt(LMS, PKs) {
	    while more rows on the LMS
	      if PK is in PKs
	        then update row
	        move to the next row
	    end while
	
	}

The main contributions of this paper are twofold: 1) to present the CTSA based on CLI and with embedded concurrency at the level of LMS; 2) to carry out a performance assessment of a case study based on a JDBC component derived from the proposed architecture. It is expected that the outcome of this paper may contribute to open a new approach to improve the performance of database applications whenever several threads need to share the same LMS instances.

Throughout this paper all examples are based on Java, SQL Server 2008 and JDBC (CLI) for SQL Server (sqljdbc4.jar). The presented source code may not execute properly, since we will only show the relevant parts for the points under discussion.

The paper is structured as follows: section 2 presents the required background; section 3 presents the related work; section 4 presents the proposed architecture and a proof of concept based on Java and JDBC; section 5 presents the performance assessment and Section 6 presents the final conclusion.

2   Background

LMS have been loosely presented and some properties have also been already described. Next follows a more detailed description about the features of LMS. LMS are instantiated by CLI to manage the data returned by Select expressions. As such, at this point it is advisable to discuss some LMS features that are relevant to this research. Fig. 1 presents a general LMS containing 5 tuples (1 to 5) and 6 attributes (a, b, c, d, e, f). This LMS could have been instantiated to manage the data returned by the following CRUD expression: *Select a, b, c, d, e, f from Table Where ....* In this case, the CRUD expression has returned 5 tuples (rows) and the current selected   tuple is row number 2. The access to LMS attributes is accomplished by selecting a tuple and then, through an index or through a label (usually the attribute name), by selecting one attribute at a time. For example, to execute an action (read, insert or update) on attribute c of tuple 2 the following steps are necessary: select tuple 2 and then execute action (index of attribute c) or action (labelof attribute c).

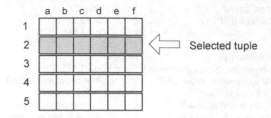

Fig. 1 LMS with 5 tuples (rows) and 6 attributes (a till f)

CLI are responsible for providing services to allow applications to scroll on LMS, to read their contents and to modify (insert, update, delete) their internal contents. Other services are also available but they are not relevant for this research. Services may be split in two categories: basic services and advanced services. Basic services comprise two groups of protocols: the scrolling protocols are aimed at scrolling on tuples and the read protocol is aimed at reading the tuples' attributes. Advanced services are available only if LMS are updatable. In this case applications are allowed to change the internal state of LMS. Advanced services comprise three protocols: insert protocol to add new tuples, update protocol to update existent in-memory tuples and, finally, delete protocol to delete existent tuples. After being committed, the new states of LMS are automatically committed into the host database. To execute any of the previous services it is necessary to know that the access to LMS is simultaneously tuple oriented and protocol oriented. This has two main implications. First, at any time only one tuple may be selected as the target tuple. Second, if a protocol is being executed, applications should not start any other protocol. If these rules ar not fulfilled, LMS may lose their previous states. For example, if an advanced service is being executed and another protocol is triggered, LMS discard all changes made during the first protocol. Table 2 concisely presents the four main protocols that are used to interact with LMS.

Read Protocol (1)
During the read protocol, attributes are read one by one and always from the current selected tuple. If a different tuple is selected, the next attribute value will be retrieved from the new selected tuple.

Update Protocol (2)
During the update protocol, attributes are updated one by one on the current selected tuple. The protocol may or may not be triggered by invoking a specific method. It ends when a specific method is invoked to commit the updated

attributes. If another tuple or protocol (except the read protocol) is selected while it is being executed, all previous changes will be discarded.

Insert Protocol (3)
The insert protocol is triggered by invoking a specific method. Then, each attribute is inserted one by one. After all attributes have been inserted, the protocol ends when a specific method is invoked to commit the inserted tuple. If another tuple or protocol (except the read protocol) is selected while it is being executed, all previous changes will be discarded.

Delete Protocol (4)
The delete protocol comprises a single method that removes the current selected tuple from the in-memory of LMS. The delete action is also automatically committed in accordance with the established policy.

Table 2 Main protocols of LMS

ID	Protocol	ID	Protocol
1	Point to a tuple Read attributes	2	Point to a tuple Start update protocol Update attributes Commit update
3	Start insert protocol Insert attributes Commit insert	4	Point to a tuple Delete tuple

3   Related Work

A research has been carried out around tools aimed at integrating client applications and databases. A survey was made for the most popular tools, such as Hibernate [8], Spring [16], TopLink [17], JPA [11] and LINQ [18]. These tools may provide concurrency but always at a very high level. Basically, they provide some locking policies implemented in order to synchronize read and write actions. But these read and write synchronized actions are not executed over the same memory location. They are executed over distinct objects, such as sessions in Hibernate. These objects (sessions) are not thread-safe and therefore do not provide any protocol to access concurrently the in-memory data.

In [19] is presented a concurrent version of the TDS protocol [20]. Unlike CTSA, the concurrency is internally implemented at the level of the TDS protocol through the services stacked above the TDS protocol. Authors have achieved significant results for the services they have implemented. Unfortunately, the research only addressed a restrict number of services not leading to a replicable and usable approach.

Aspect-oriented programming [21] community considers persistence as a crosscutting concern [22]. Several works have been presented but none addresses the points here under consideration. The following works are emphasized: [23] is focused on separating scattered and tangled code in advanced transaction management; [22] addresses persistence relying on AspectJ; [24] presents AO4Sql as an aspect-oriented extension for SQL aimed at addressing logging, profiling and runtime schema evolution. It would be interesting to see an aspect-oriented approach for the points herein under discussion.

In [25] a different approach is presented to address the lack of concurrent mechanisms of CLI. Currency is implemented by an explicit locking mechanism based on two methods: *lock()* and *unlock*. Programmers are responsible for invoking these methods correctly in order to control the exclusive access mode to LMS. Additionally, the conducted assessment is based on a fixed number of rows which does not convey a dynamic perspective of the performance for different scenarios.

To the best of our knowledge no other researches have been conducted around concurrency on LMS of CLI.

4   Concurrent Tuple Set Architecture

In this section we start to present CTSA and then a proof of concept is also presented.

4.1   CTSA Presentation

CTSA defines the concept of *execution context* as the information needed to characterize, at any time, the interaction between a thread and a component based on the CTSA. The execution context of each thread comprises the protocol that is being executed and the current selected tuple. This concept is very important because it is the basis for the concurrent implementation of LMS. In concurrent environments, each thread must have a complete control on the tuple and on the protocol it is executing. If this is not ensured, a running thread may be preempted by another thread that changes the execution context. The first thread will never be aware about this situation and when it becomes the running thread it will execute its actions in a different execution context. In order to keep full control on the execution context, each thread needs to access the LMS in exclusive mode and also to be able to assure that it runs on its own execution context. The former condition ensures that other threads are not allowed to change the execution context of protocols that are being executed. The latter condition ensures that at the beginning of any protocol, if necessary, every thread is able to restore its execution context. To decide upon which strategy to follow to implement both conditions, two possibilities were considered and tested: 1) method oriented: execution context is managed method by method; 2) protocol oriented: execution context is managed at the protocol level. Table 3 briefly shows the logic associated with each approach. The scrolling process involves one method at a time and, therefore, it is implemented as method oriented access mode. Access modes for Insert,

Update and Delete protocols do not have any other alternative but be implemented as protocol oriented. This derives from the fact, as mentioned before, that these protocols cannot be preempted to start a different protocol. Read protocol may be implemented in any access mode. To decide upon which access mode to implement some tests with the two access modes were carried out. The collected results have shown, for the same scenarios, that performance and concurrency improvement depend on the same variable but in opposite ways. They depend on the number of times that threads are preempted by other threads. Every time this occurs, a change in the execution contexts must be performed. When this number increases, performance tends to decrease and concurrency tends to increase. When this number decreases, performance tends to increase and concurrency tends to decrease. Thus, in order to improve performance, it was decided to implement the Read protocol based on the protocol oriented access mode.

Table 3 Approaches for the exclusive access mode to LMS

Method Oriented	Protocol Oriented
1. get exclusive access	1. get exclusive access
2. set execution context	2. set execution context
3. execute method	3. while protocol is not over execute method
4. store execution context	4. store execution context
5. release exclusive access	5. release exclusive access

«interface» IRead	«interface» IInsert
+beginRead() +endRead() +getInt(in idx : long(idl)) : long(idl) +getString(in idx : long(idl)) : string(idl) +...()	+beginInsert() +endInsert() +cancelInsert() +setInt(in idx : long(idl), in value : string(idl)) +setString(in idx : long(idl), in value : string(idl)) +...()

«interface» IScroll	
+moveNext() : bool +moveFirst() : bool +moveAbsolute(in position : int) : bool +isFirst() : bool +...()	

	«interface» IUpdate
	+beginUpdate() +endUpdate() +cancelUpdate() +setInt(in idx : long(idl), in value : string(idl)) +setString(in idx : long(idl), in value : string(idl)) +...()

«interface» IDelete
+delete()

Fig. 2 CTSA main protocols

Fig. 2 presents the interfaces for the five main protocols: *IRead* (read protocol), *IInsert* (insert protocol), *IUpdate* (update protocol), *IDelete* (delete protocol) and IScroll (scroll protocol). Only the main methods of IRead, IUpdate, IInsert and IScroll have been presented in order not to overcrowd the class diagrams. Exclusive access modes based on the protocol oriented strategy are started by the execution of an explicit starting method (*beginRead, beginUpdate* and *beginInsert*) and released only after the execution of another explicit method (*endRead, endUpdate* and *endInsert*). This strategy ensures the exclusive access to LMS while the protocol is being executed and also the initialization of the correct execution context before any access to the LMS. *getInt* and *getString* methods read attributes (read protocol) of types integer and string, respectively, from LMS. *setInt* and *setString* methods set the values for the attributes (Update and Insert protocols) of type integer and string, respectively. Beyond these methods (get and set), there are other methods each one suited to deal with one data type of the host programming language. Exclusive access mode of IScroll methods and IDelete method are method oriented and, therefore, no additional methods are needed.

Fig. 3 presents a simplified CTSA class diagram. Concurrent threads sharing the same LMS receive a new CTSA instance where all CTSA instances share the same LMS. *lms* is the LMS instance, *currentTuple* (current selected tuple) and *protocol* (protocol being used) define the execution context of the owner thread, *setExecution* restores the execution context of the running thread and *storeExecution* stores the current execution context.

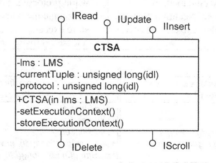

Fig. 3 CTSA class diagram

4.2   *Proof of Concept*

This section evaluates CTSA using a proof of concept implemented in Java and JDBC and it uses the *ReentrantLock* [26] synchronization entity to guarantee exclusive access to shared data structures when threads interact with components based on the CTSA. Any other java mechanism, such as *synchronized methods* [27], could have been used. Due to space limitations we will only present CTSA from users' perspective, see Fig. 4. The thread receives a CTSA instance (line 23). When the thread enters the running state (line 27), it iterates the LMS one tuple at a time (line 30). This access mode is method oriented and, as such, there is no starting trigger. The tuple is read (line 32-33). The access mode of the read protocol is protocol oriented and, therefore, there is a trigger to start (line 31) and a trigger to stop it (line 34). This example shows that users of components based on the CTSA have the advantage of using thread safe LMS without any concern about its implementation. The use of the remaining protocols is very similar to the ones presented in Fig. 4 and, therefore, no additional examples are needed.

```
23 User( CTSA ctsa ) {
24       this.ctsa = ctsa;
25 }
   @Override
27 public void run() {
28       // ... code
29       try {
30           while ( ctsa.moveNext() ) {
31               ctsa.beginRead();
32               id = ctsa.getInt( 1 );
33               // ... read other attributes
34               ctsa.endRead();
35               // other protocols
36           }
37       } catch ( SQLException ex ) {}
38 }
```

Fig. 4 CTSA from users' perspective

5   CTSA Assessment

Performance assessment was carried out comparing two entities known as the Component CTSA (C-CTSA) and the Concurrent JDBC (C-JDBC). C-CTSA is responsible for evaluating components relying on the CTSA architecture and it is based on a component derived from the proof of concept here presented. C-JDBC is responsible for evaluating a concurrent approach based on the standard JDBC. The evaluation of both entities comprises a single façade: performance. Three scenarios were defined for both components: Select (s), Update (u) and Insert (i). Each scenario comprises a set of several numbers of tuples to be processed [nr] and a set of several numbers of simultaneous running threads [nt]. In order to formalize the entities' representation we define $E(\alpha,p,\gamma)$ ([nt], [nr]) where $\alpha \in \{$c-ctsa,c-jdbc$\}$, p is for performance façade and $\gamma \in \{$s,u,i$\}$. To simplify, $E(\alpha,p,\gamma)$ ([nt], [nr]) is represented by default as $E(\alpha,p,\gamma)$. Each scenario comprises a specific goal which is known as a task. A task represents a particular case for the use of C-CTSA and C-JDBC regarding the LMS. The tasks to be performed are: Read (read [nr] tuples from the LMS), Update (update [nr] tuples of a LMS) and Insert (insert [nr] tuples into a LMS). It was decided to create a favorable environment to C-JDBC and an unfavorable environment for C-CTSA to execute the defined tasks. This way, the minimum performance of real scenarios based on C-CTSA should be delimited by the collected measurements. This issue will be addressed in more detail after explaining the SQL Server behavior about LMS.

The test-bed comprises two computers: PC1 - Dell Latitude E5500, Intel Duo Core P8600 @2.40GHz, 4.00 GB RAM, Windows Vista Enterprise Service Pack 2 (32bits), Java SE 6, JDBC (sqljdbc4); PC2 – Asus-P5K-VM, Intel Duo Core

E6550 @2,33 GHz, 4.00 GB RAM, Windows XP Professional Service Pack 3, SQL Server 2008. C-JDBC and C-CTSA are executed in PC1 and SQL Server runs in PC2. In order to promote an ideal environment the following actions were taken: the running threads were given the highest priority and all non-essential processes/services were cancelled in both PCs; a direct and dedicated network cable connecting PC1 and PC2 has been used in exclusive mode and performing 100MBits of bandwidth. Transactions were not used and auto-Commit has been always enabled. A new database was created in conformance with the schema presented in Fig. 5 to assess both entities. In order to avoid any overhead added by SQL Server, some default SQL Server database properties were changed as, Auto Update Statistics = false and Recovery Model = Simple.

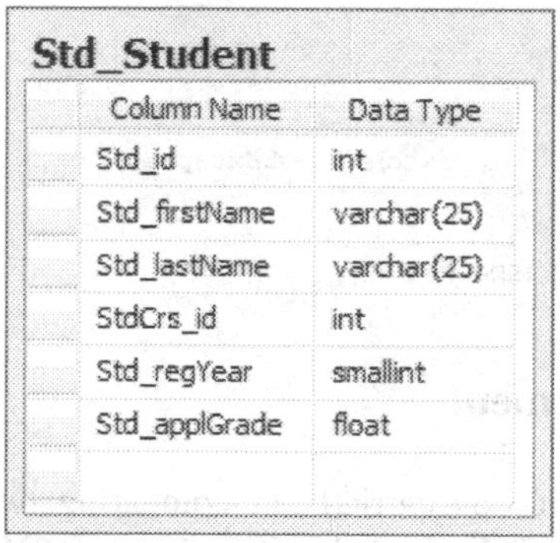

Std_Student

Column Name	Data Type
Std_id	int
Std_firstName	varchar{25}
Std_lastName	varchar{25}
StdCrs_id	int
Std_regYear	smallint
Std_applGrade	float

Fig. 5 Std_Student schema

Some important aspects are out of the scope of this study. Aspects as database server performance, network delays and memory consumption are not individually addressed but considered as part of the overall environment. This has been assumed because both entities share the same infrastructure.

It is essential to have some knowledge about SQL Server behavior, which is similar to most of the other relevant relational database management systems, to completely understand the details of each defined task and also to understand the collected results. When a Select statement is executed using a scrollable or an updatable LMS, SQL Server creates a server cursor with all the selected tuples. These tuples are dynamically transferred in blocks, from the server to LMS, whenever necessary. This means that at any time LMS may not have all the tuples but only a sub-set of all tuples. When users point to a tuple that is not present in the LMS, the TDS protocol discards the current LMS content and fetches the block containing the desired tuple. This has a deep implication. If threads are always requesting tuples that are not present in the LMS, SQL Server has to transfer the correspondent block for each request. In an extreme scenario, each individual action over the LMS could imply a new transference of tuples. From the previous statements, it is expected that the number of blocks to be transferred will increase when the number of tuples, inside server cursors, increases and also when the dispersion of the used policy to select tuples, contained by server cursors, increases. Thus, to create the environments for both entities, the following decisions were taken:

C-JDBC (favorable environment) - each thread will always access tuples sequentially from the first one till the last one.

C-CTSA (unfavorable environment) - two conditions were implemented: 1) after accessing a tuple, each thread will give the opportunity for other threads to become the running thread (this will maximize the number of changes in the execution context); 2) each thread will have its own set of tuples, not shared with any other thread (this will maximize the number of blocks of tuples to be transferred from the server cursor to LMS).

Table 4 Algoritm for the $E_{(c\text{-}ctsa,p,\gamma)}$ assessment

```
1. Delete all rows from Std_Student
2. Fill Std_Student with [nr]*[nt] rows (zero rows for insert)
3. Start counter
4. Select all rows from Std_Student into one single ResultSet
5. Create all threads. Each thread (ψ tuples)
       5.1 for each tuple
              5.1.1 read/update/insert (tuple)
              5.1.2 suspend thread
       5.2 dies
6. Wait all threads to die
7. Stop counter
```

Table 5 Algoritm for the $E_{(c\text{-}jdbc,p,\gamma)}$ assessment

```
1. Delete all rows from Std_Student
2. Fill Std_Student with [nr]*[nt] rows (zero rows for insert)
3. Start counter
4. Create all threads. Each thread:
       4.1  select ψ tuples into its own ResultSet
       4.2  for each tuple
              4.2.1 read/update/insert a tuple
       4.3  dies
5. Wait all threads to die
6. Stop counter
```

Table 4 shows the algorithm for the assessment of E(c-ctsa,p,γ). The same ResultSet (LMS) is shared by all [nt] threads. Each thread executes its scenario for a group of ψ=[nr] adjacent tuples and auto-suspends itself after accessing each tuple. The intersection of all $\psi=\varnothing$.

Table 5 shows the algorithm for the assessment of $E_{(c\text{-}jdbc,p,\gamma)}$. Each thread creates its own ResultSet (LMS) containing/inserting a group of ψ=[nr] adjacent tuples. The intersection of all $\psi=\varnothing$.

To contextualize the performance assessment environment some initial measurements were carried out to delimit the range of [nt] and [nr] to be used. In order to emphasize concurrency mechanisms, priority was given to the range of [nt] in detri-ment of [nr]. Values for these metrics were collected by empirical experi-mentation based on an iterative process. The idea is to gather a set of values for [nt] and [nr] that may be used to assess and compare the performance of both $E(\alpha,p,\gamma)$ entities. To accomplish this, both entities, $E(c\text{-}ctsa,p,\gamma)$ and $E(c\text{-}jdbc,p,\gamma)$, were executed under several combinations of [nt] and [nr] until the collected val-ues comprise a range of behaviors considered satisfactory to accurately assess and compare the performance of both entities. After several iterations it was decided that a reliable execution environment should be defined as:

$$[nt]=\{1, 5, 10, 25, 50, 75, 100, 150, 200, 250, 350, 500\}$$
$$[nr]=\{5, 10, 25, 50, 75, 100\}$$

In accordance with the requirements, this execution environment evaluates the performance by maximizing the number of simultaneous running threads in detri-ment of the number of tuples. With 500 threads and 100 tuples it was possible to accurately assess and foresee the performance of both entities. This was the main reason for their acceptance. The intermediate collected values showed to be enough to obtain well defined charts for the behaviors of both entities. Just as a final note, some scenarios took some minutes to setup and to process the highest values of [nt] and [nr]. This knowledge was also considered to delimit the two top values (nr=100 and nt=500), this way avoiding any risk to successfully accom-plish the collecting process of all necessary measurements. 25 raw measures were collected for each $E(\alpha,p,\gamma)([nt],[nr])$ leading to $(2\times3\times12\times6)\times25=10,800$ raw mea-surements. Intermediate measurements were computed from the average of the 5 best measures of each $E(\alpha,p,\gamma)([nt],[nr])$ leading to a total of $2\times3\times12\times6=432$ mea-surements. The final measurements used in the next charts represent the ratios between $E(c\text{-}jdbc,p,\gamma)$ and $E(c\text{-}ctsa,p,\gamma)$ for each $([nt],[nr])$. In all charts the vertic-al axis is for the ratios and the horizontal axis is for the [nt].

Select Scenario: The chart for the select scenario is shown in Fig. 6. From it, it is clear that the ratios decrease whenever the number of tuples increases and when-ever the number of threads increases. $E(c\text{-}jdbc,p,\gamma)$ have [nt] server cursors and each thread sequentially reads its own tuples from the first one till the last one. Thus, the transference of block of tuples only happens when a thread tries to read the next tuple that is after the last one contained in the ResultSet. Moreover, for this entity different threads do not compete for the same ResultSet this way avoid-ing any randomness in the tuples to be selected. Regarding $E(c\text{-}ctsa,p,\gamma)$ there is only one server cursor shared by all threads. The implemented Read task signifi-cantly increases the possibility of each thread to be requesting a tuple that is not present in the ResultSet and, therefore, to trigger a new transference of block of tuples. With other strategies where threads read shared sets of tuples, the block transference rate should be much lower, this way increasing the ratios between the two entities. Another relevant issue is that the Select scenario is a light scenario mainly because the Select expression and the read protocol are very efficient when compared with the other CRUD expressions and other protocols. Thus, the overhead induced by the blocks transference have a deeper impact in the overall performance. The impact increases with the number of tuples and the number of threads as expected. Anyway, the collected results show that for lower values of number of tuples and lower values of number of threads the ratios vary between 1.02 and 3.44 times, as shown in Fig. 7 (the cells with a rose color contain ratios below one). It may also be seen that the worst ratio achieves 0.80 for nt=100 and nr=25. These results show that despite the unfavorable test conditions for C-CTSA, C-CTSA still achieves significant results. For example, the relative highest gain in performance (3.44) is much more significant than the relative highest lost in performance (0.8). Fig. 7 presents a detailed vision for the ratio between both entities for all combinations of [nr] and [nt].

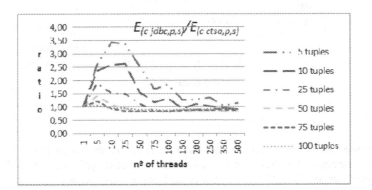

Fig. 6 $E_{(c\text{-}jdbc,p,s)} / E_{(c\text{-}ctsa,p,s)}$ chart

NR/NT	5	10	25	50	75	100
1	1,03	1,04	1,03	1,03	1,02	1,05
5	2,61	2,38	1,90	1,41	1,22	1,07
10	3,44	2,60	1,53	1,09	0,93	0,96
25	3,38	2,63	1,51	0,94	0,86	0,97
50	2,54	1,57	1,14	0,83	0,83	0,89
75	1,69	1,18	0,85	0,84	0,83	0,89
100	1,86	1,34	0,80	0,86	0,84	0,88
150	1,28	0,95	0,87	0,84	0,86	0,91
200	1,28	1,12	0,90	0,87	0,88	0,92
250	1,35	1,02	0,92	0,87	0,87	0,92
350	1,06	0,95	0,94	0,86	0,89	0,93
500	1,16	0,92	0,93	0,86	0,87	0,94

Fig. 7 $E_{(c\text{-}jdbc,p,s)} / E_{(c\text{-}ctsa,p,s)}$ details

Update Scenario: The chart for the update scenario is shown in Fig. 8. The comments made to the Select scenario are also applied to the Update scenario regarding the transference rate of block of tuples. The most significant differences are: 1) the update protocol is a heavy protocol and, thus, its overhead has a deep impact on both entities and in the collected measurements; 2) the $E_{(c\text{-}jdbc,p,\gamma)}$ entity has *[nr]* server cursors, each one competing with the others to update the same requested attributes while $E_{(c\text{-}ctsa,p,\gamma)}$ entity has only one server cursor and the competition is performed at the client side. Despite the unfavorable conditions for C-CTSA, in this scenario, the ratio is always significantly greater than 1. It increases in the range 1<*nt*<10 and for *nt*>10 the ratios are practically stable for each individual *[nr]* (except for *nr*=5). Another relevant issue is that the ratios decrease when *[nr]* increases for every *[nt]*.

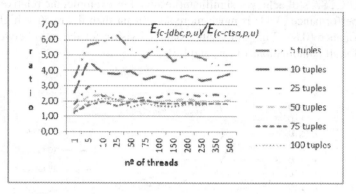

Fig. 8 $E_{(c\text{-}jdbc,p,u)} / E_{(c\text{-}ctsa,p,u)}$ chart

Insert Scenario: The chart for the insert scenario is shown in Fig. 9. The most relevant aspect is the slight but constant ratios increase with *[nt]* for each *[nr]*. In the initial stage, ResultSets are empty and tuples are sequentially inserted and committed one by one in the host database table. In this scenario, in opposite to the others, all $E_{(c\text{-}ctsa,p,\gamma)}$ threads insert adjacent tuples this way minimizing the number of blocks to be transferred. In spite of being a very heavy scenario for both entities, the differences between C-CSTA and C-JDBC are enough to be noticed in the ratios. It is always greater than 1 and higher values of *[nr]* cause a decreasing in the ratios.

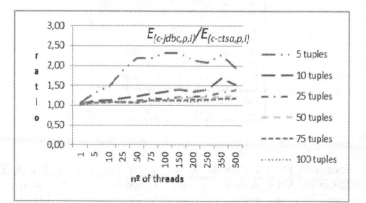

Fig. 9 $E_{(c\text{-}jdbc,p,i)} / E_{(c\text{-}ctsa,p,i)}$ chart

6   Conclusions

In this paper an architecture for a concurrent LMS, herein known as CTSA, has been presented. A proof of concept has also been presented based on a standard JDBC API. In order to assess CTSA performance in a concurrent environment and compare it with an equivalent environment based on a standard JDBC solution, a test-bed has been defined and implemented with two concurrent entities: C-JDBC and C-CTSA. C-CTSA was assessed in unfavorable conditions and C-JDBC has been assessed in favorable conditions in order to delimit and evaluate C-CTSA performance minimum gain. In spite of these conditions, C-CTSA always gets better scores for the update and for the insert scenarios. In the Select scenario, C-CTSA obtained significant scores in the range of lower values of *[nr]* and *[nt]*. Anyway, for higher values of *[nr]* and *[nt]* the minimum ratio did not go below 0.8 which is still a remarkable score.

The outcome of this research should encourage CLI providers to release CLI with internal embedded concurrency. Embedded concurrency has the advantage of accessing the LMS's internal data structures to optimize the implementation of the different protocols.

References

1. David, M.: Representing database programs as objects. In: Bancilhon, F., Buneman, P. (eds.) Advances in Database Programming Languages, pp. 377–386. ACM, N.Y. (1990)
2. Cook, W., Ibrahim, A.: Integrating programming languages and databases: what is the problem? ODBMS.ORG, Expert Article (2005)
3. Eisenberg, A., Melton, J.: Part 1: SQL Routines using the Java (TM) Programming Language. In: American National Standard for Information for Technology Database Languages - SQLJ, International Committee for Information Technolgy (1999)
4. Kulkarni, D., et al.: LINQ to SQL: .NET Language-Integrated Query for Relational Data. Microsoft
5. ISO. ISO/IEC 9075-3:2003 (2003),
   `http://www.iso.org/iso/catalogue_detail.htm?csnumber=34134`
   (May 2011)
6. Parsian, M.: JDBC Recipes: A Problem-Solution Approach. Apress, NY (2005)
7. Microsoft. Microsoft Open Database Connectivity (1992),
   `http://msdn.microsoft.com/en-us/library/`
   `ms710252(VS.85).aspx` (July 2012)
8. Christian, B., Gavin, K.: Hibernate in Action. Manning Publications Co. (2004)
9. Oracle. Oracle TopLink (October 2011),
   `http://www.oracle.com/technetwork/middleware/toplink/overv`
   `iew/index.html`
10. Oracle. Java Data Objects (JDO) (November 2011),
    `http://www.oracle.com/technetwork/java/`
    `index-jsp-135919.html`
11. Yang, D.: Java Persistence with JPA. Outskirts Press (2010)

12. IBM. Introduction to Service Data Objects (November 2011),
    http://www.ibm.com/developerworks/java/library/j-sdo/
13. Mead, G., Boehm, A.: ADO.NET 4 Database Programming with C# 2010. Mike Mu-
    rach & Associates, Inc., USA (2011)
14. Oracle. ResultSet (July 2012),
    http://docs.oracle.com/javase/6/docs/api/java/sql/
    ResultSet.html
15. Microsoft. RecordSet (ODBC) (June 2012),
    http://msdn.microsoft.com/en-us/library/5sbfs6f1.aspx
16. Spring. Spring (November 2011), http://www.springsource.org/
17. Oracle Database (May 2010),
    http://www.oracle.com/us/products/database/index.html
18. Erik, M., Brian, B., Gavin, B.: LINQ: Reconciling Object, Relations and XML in the
    .NET framework. In: ACM SIGMOD Intl. Conf. on Management of Data. ACM, Chi-
    cago (2006)
19. Gomes, D., Pereira, Ó.M., Santos, W.: JDBC (Java DB connectivity) concorrente. In:
    DETI, p. 115. University of Aveiro: ria - institutional repository (2011),
    http://hdl.handle.net/10773/7359
20. Microsoft. [MS-TDS]: Tabular Data Stream Protocol Specification (July 2012),
    http://msdn.microsoft.com/en-
    us/library/dd304523(v=prot.13).aspx
21. Akşit, M., Tekinerdoğan, B.: Aspect-Oriented Programming Using Composition-
    Filters. In: Demeyer, S., Dannenberg, R.B. (eds.) ECOOP 1998 Workshops. LNCS,
    vol. 1543, pp. 435–435. Springer, Heidelberg (1998)
22. Laddad, R.: AspectJ in Action: Practical Aspect-Oriented Programming. Manning
    Publications, Greenwich (2003)
23. Fabry, J., D'Hondt, T.: KALA: Kernel Aspect Language for Advanced Transactions.
    In: Proceedings of the 2006 ACM Symposium on Applied Computing. ACM, Dijon
    (2006)
24. Dinkelaker, T.: AO4SQL: Towards an Aspect-Oriented Extension for SQL. In: 8th
    Workshop on Reflection, AOP and Meta-Data for Software Evolution (RAM-SE
    2011), Zurich, Switzerland (2011)
25. Pereira, O.M., Aguiar, R.L., Santos, M.Y.: Assessment of a Enhanced ResultSet Com-
    ponent for Accessing Relational Databases. In: ICSTE-Int. Conf. on Software Tech-
    nology and Engineering, Puerto Rico (2010)
26. Orcale. Call ReentrantLock (November 2012),
    http://docs.oracle.com/javase/6/docs/api/java/util/
    concurrent/locks/ReentrantLock.html
27. Oracle. Synchronized Methods (November 2012),
    http://docs.oracle.com/javase/tutorial/essential/
    concurrency/syncmeth.html

Analog Learning Neural Network Using Two-Stage Mode by Multiple and Sample Hold Circuits

Masashi Kawaguchi, Naohiro Ishii, and Masayoshi Umeno

Abstract. In the neural network field, many application models have been proposed. A neuro chip and an artificial retina chip are developed to comprise the neural network model and simulate the biomedical vision system. Previous analog neural network models were composed of the operational amplifier and fixed resistance. It is difficult to change the connection coefficient. In this study, we used analog electronic multiple and sample hold circuits. The connecting weights describe the input voltage. It is easy to change the connection coefficient. This model works only on analog electronic circuits. It can finish the learning process in a very short time and this model will enable more flexible learning.

Key words: Electronic circuit, neural network, multiple circuit.

1 Introduction

We propose the dynamic learning of the neural network by analog electronic circuits. This model will develop a new signal device with the analog neural

Masashi Kawaguchi
Department of Electrical & Electronic Engineering,
Suzuka National College of Technology, Shiroko, Suzuka Mie Japan
e-mail: masashi@elec.suzuka-ct.ac.jp

Naohiro Ishii
Department of Information Science, Aichi Institute of Technology,
Yachigusa, Yagusa-cho, Toyota, Japan
e-mail: ishii@aitech.ac.jp

Masayoshi Umeno
Department of Electronic Engineering, Chubu University, 1200 Matsumoto-cho,
Kasugai, Aichi 487-8501 Japan
e-mail: umeno@solan.chubu.ac.jp

R. Lee (Ed.): *Computer and Information Science*, SCI 493, pp. 159–170.
DOI: 10.1007/978-3-319-00804-2_12   © Springer International Publishing Switzerland 2013

electronic circuit. One of the targets of this research is the modeling of biomedical neural function. In the field of neural network, many application models have been proposed. And there are many hardware models that have been realized. These analog neural network models were composed of the operational amplifier and fixed resistance. It is difficult to change the connection coefficient.

1.1   Analog Neural Network

The analog neural network expresses the voltage, current or charge by a continuous quantity. The main merit is it can construct a continuous time system as well as a discrete time system by the clock operation. Obviously, the operation of the actual neuron cell utilizes analog. It is suitable to use an analog method for imitating the operation of an actual neuron cell. Many Artificial neural networks LSI were designed by the analog method. Many processing units can be installed on a single-chip, because each unit can be achieved with a small number of elements, addition, multiplication, and the nonlinear transformation. And it is possible to operate using the super parallel calculation. As a result, the high-speed offers an advantage compared to the digital neural network method [1][2]. In the pure analog circuit, the main problem is the achievement of an analog memory, how to memorize analog quantity [3].This problem has not been solved yet. The DRAM method memorizes in the capacitor as temporary memory, because it can be achieved in the general-purpose CMOS process [4]. However, when the data value keeps for a long term, digital memory will also be needed. In this case, D/A and A/D conversion causes an overhead problem. Other memorizing methods are the floatage gate type device, ferroelectric memory (FeRAM) and magnetic substance memories (MRAM) [5][6].

1.2   Pulsed Neural Network

Another hardware neural network model has been proposed. It uses a pulsed neural Network. Especially, when processing time series data, pulsed neural network model has good advantages. In particular, this network can keep the connecting weights after the learning process [7]. Moreover, the reason the learning circuit used the capacitor is that it takes a long time to work the circuits. In general, the pulse interval of the pulsed neural network is about 10μS. The pulsed neuron model represents the output value by the probability of neuron fires. For example, if the neuron is fired 50 times in a 100 pulse interval, the output value is 0.5 at this time. To represent the analog quantity using the Pulsed Neuron Model, it needs about 100 pulses. Thus, about 1mS is needed to represent the output analog signal on a pulsed neuron model.

In this study, we used the multiple circuits. The connecting weights describe the input voltage. It is easy to change the connection coefficient. This model works only on analog electronic circuits. It can finish the learning process in a very short time and this model will allow for more flexible learning. Recently, many researchers have focused on the semiconductor integration industry. Especially, low electrical power, low price, and large scale models are important. The neural network model explains the biomedical neural system. Neural network has flexible learning ability. Many researchers simulated the structure of the biomedical brain neuron using an electronic circuit and software.

1.3 Overview

The results of the neural network research provide feedback to the neuro science fields. These research fields were developed widely. The learning ability of a neural network is similar to the human mechanism. As a result, it is possible to make a better information processing system, matching both advantages of the computer model and biomedical brain model. The structure of the neural network usually consists of three layers, the input layer, intermediate layer and output layer. Each layer is composed of the connecting weight and unit. A neural network is composed of those three layers by combining the neuron structures [8][9].

In the field of neural network, many application methods and hardware models have been proposed. A neuro chip and an artificial retina chip are developed to comprise the neural network model and simulate the biomedical vision system. In this research, we are adding the circuit of the operational amplifier. The connecting weight shows the input voltage of adding circuits. In the previous hardware models of neural net-work, changing connected weights was difficult, because these models used the resistance elements as the connecting weights.

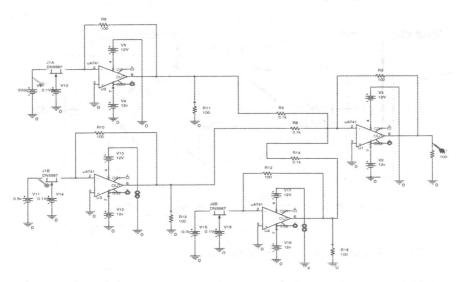

Fig. 1 Neural Circuit (Two-input and One-output)

Moreover, the model which used the capacitor as the connecting weights was pro-posed. However, it is difficult to adjust the connecting weights. In the present study, we proposed a neural network using analog multiple circuits. The connecting weights are shown as a voltage of multiple circuits. It can change the connecting weights easily. The learning process will be quicker. At first we made a neural network by computer program and neural circuit by SPICE simulation. SPICE means the Electric circuit simulator as shown in the next chapter. Next we measured the behavior confirmation of the computer calculation and SPICE simulation. We compared both output results and confirmed some extent of EX-OR behavior [10].

2   SPICE

In this research, we used the electric circuit simulator SPICE. Electric circuit simulator (SPICE) is the abbreviation of Simulation Program with Integrated Circle Emphasis. It can reproduce the analog operation of an electrical circuit and the electric circuit. After this, the circuit drawn by CAD, set the input voltage. SPICE has the function of AC, DC and transient analysis. At first, we made the differential amplifier circuits and Gilbert multipliers circuits. And we confirmed the range of voltage operated excellently. The neuron structure was composed of multiple circuits by an operational amplifier for multiplication function achievement, current mirror circuits to achieve nonlinear function and differential amplifier circuits.

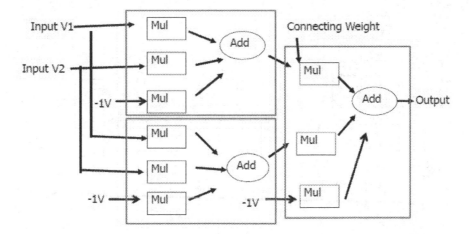

Fig. 2 The Architecture of Three-Layers Neural Circuits

In the previous hardware model of neural network, we used the resistance element as a connecting weight. However, it is difficult to change the resistance value. In the neural connection, it calculates the product the input value and connecting weight. We used the multiple circuit as the connecting weight. Each two inputs of multiple circuits means an input value and connecting weight. The connecting weight shows the voltage value. It is easy to change the value in the learning stage of neural network.

Figure 1 is the neural circuit of two inputs and one output which reproduces the characteristic of one neuron, using current addition by current mirror circuits, the product of the input signal and connecting weights.

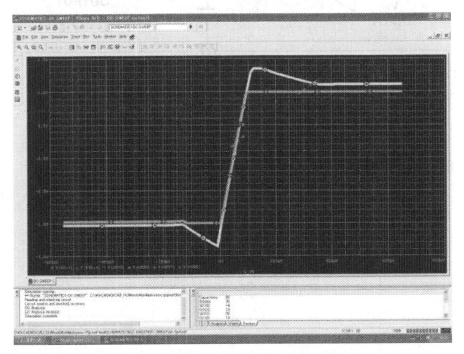

Fig. 3 Experimental Result of Three-Layers Neural Circuits

3   Three Layers Neural Network

We constructed a three layer neural network, an input layer, middle layers and an output layer. There are two input units, two middle units and one output unit. We combined the neural unit described in the preceding chapter. In Figure 2, we show the block diagram of a general neural network model. However it uses the multiple circuit for easy changing of the connecting weight. "Mul" means multiple circuits and "Add" means addition of circuits in Fig. 2. The experimental result is shown in Fig. 3. We confirmed when the range of the voltage is between -0.05V and 0.15V, this circuit operated normally. The linear graph is the output of the middle layer and the nonlinear graph is the output of the final layer in Fig. 3 [11]. In the middle layer, we achieved a good output signal. In the output layer, there was a little distortion signal. However, this will not present a significant problem on the neural network output.

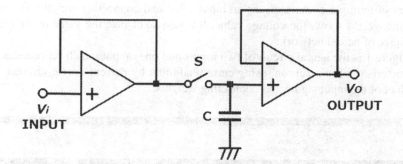

Fig. 4 Sample Hold Circuits

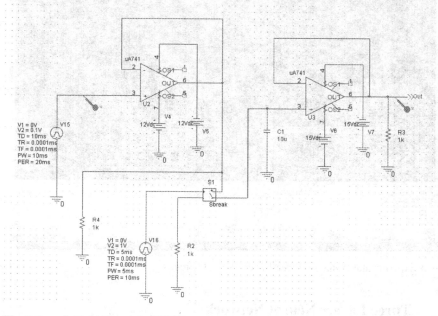

Fig. 5 Sample Hold Circuit by SPICE

4   Dynamical Learning Model

We propose the dynamical learning model using a pure analog electronic circuit. We used analog neural network, explained in a previous chapter. In the learning stage, we used analog feedback circuits.

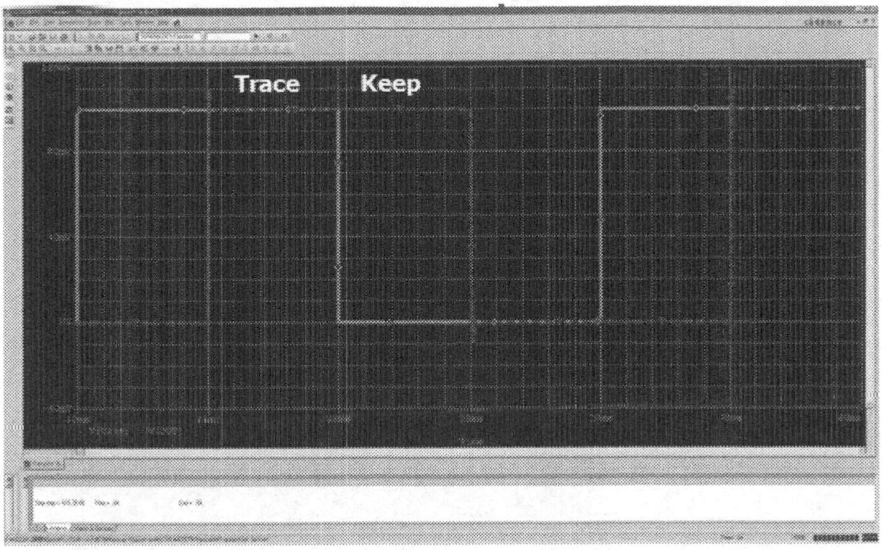

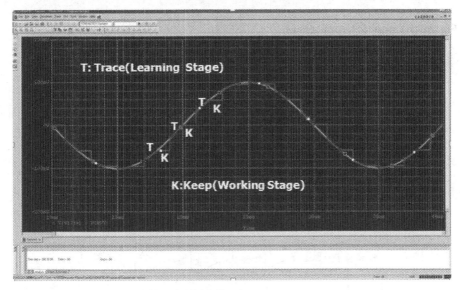

Fig. 6 Sample Hold Circuit by SPICE

Fig. 7 Simulation Result of Sample Hold Circuit

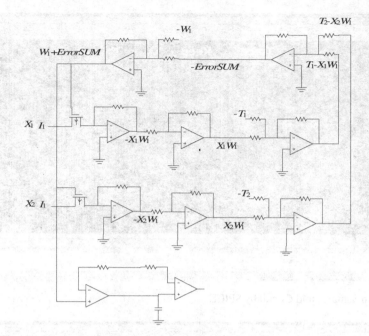

Fig. 8 The Circuit of Learning Stage

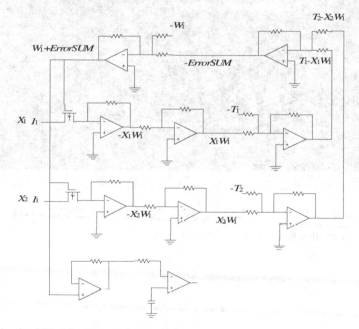

Fig. 9 The Circuit of Working Stage

We use a separate neural network of each teaching signal. Real time learning is possible. We used the sample hold circuit in the working stage. It can hold the connection weights. In the working stage, this neural network is working. This circuit can perform periodical work, learning mode and working mode [12]. In Fig. 4, we show the Sample Hold Circuits. They can keep the output value for a brief time in the holding mode when the switch "S" is turn off. However, when the switch is turn on, this circuit situation is in "sampling mode". In the sampling mode, it is the same value for the input signal and output signal.

We constructed the Sample Hold Circuit by CAD and simulated by SPICE. We show the experimental Sample Hold Circuit in Fig. 5 and confirm experimental result in Fig. 6. In Fig. 7, we show the simulation result of sample hold circuits. It represents the learning stage and working stage of Neural Circuit. We show each stage in Fig.8 and Fig. 9.

Fig. 8 shows the circuit of Learning Stage. The sample hold circuit is in the sampling mode. Fig. 6 shows the circuit of the working stage. The sample hold circuit is in holding mode. In the base of our previous paper [12], we have the additional experiment. We sated each resistance or capacitor value on the Capture CAD by SPICE, in Fig. 10. In Fig. 11, we show the result when the input signal is a square wave. We got the result, the learning time is about 20μs. After spending 20μs, the output value is constant. We assume that the working time is also 20μs. The learning cycle of this circuit is 25,000 times per second.

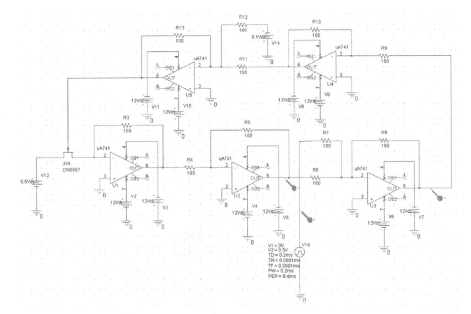

Fig. 10 The Learning Neural Circuit on Capture CAD by SPICE

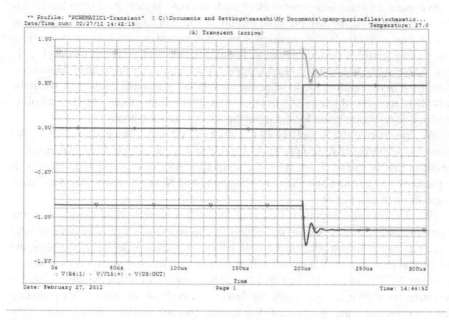

Fig. 11 The Simulation Result, input Square Wave

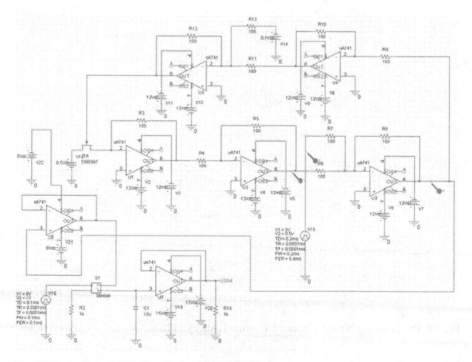

Fig. 12 Basic Neural Circuit with Sample Hold Circuits

The learning speed of this model is very high in spite of a very simple circuit using low cost elements. Repeating the learning mode and working mode, the circuits can realize flexible learning. We show the Basic Neural Circuit with Sample Hold Circuits in Fig. 12.

On the other hand, the pulsed neural Network has an advantage. Particularly, this network can also keep the connecting weights after the learning process. However, it takes a long time for the learning process when many pulses are required. As the typical pulsed neuron model, about 1000 pulses were required for the learning process. However, our proposed model is constructed with a cheap electrical device. If we use the high quality analog electrical device, the learning speed will be improved more than pulsed neuron model. In the result of this experiment the performance is low because of using general-purpose, inexpensive parts. The operating speed will be improved by using a high-performance element which has a good slew rate. However, this system is a simple circuit. The number of parts is few. The cost will not rise much even if good performance parts are used.

5   Conclusion

We constructed a three layer neural network, two-input layers, two-middle layers and one output layer. We confirmed the operation of the three layer analog neural network with the multiplying circuit by SPICE simulation.

The connection weight can change by controlling the input voltage. This model has extremely high flexibility characteristics. When the analog neural network is operated, the synapse weight is especially important. It is how to give the synapse weight to this neural network. To solve this problem, it is necessary to apply the method of the back propagation rule that is a general learning rule for the multiple electronic circuits. This neural circuit model is possible the learning. The learning speed will be rapid. And dynamic learning will be realized. The method is calculating the difference between the output voltage and the teaching signal of the different circuits and the feedback of the difference value for changing connecting weights. The learning cycle of this circuit is 25,000 times per second. The learning speed of this model is very high in spite of a very simple circuit using low cost elements.

The learning time of this model is very short and the working time of this model is almost real-time. The pulsed neuron model represents the output value by the probability of neuron fires. To represent the analog quantity using the Pulsed Neuron Model, enough time for at least a few dozen pulses is needed. The output value of this model is the output voltage of this circuit. We don't need to convert the data, we can use the raw data from this model. This model allows for switching the working mode and learning mode. It is always necessary to input the teaching signal. However, the connecting weight changes according to the changing the teaching signal. This model can also easily accommodate changes in the environment. In each scene, optimal learning is possible.

It will improve the artificial intelligence element with self dynamical learning. The realization of an integration device will enable the number of elements to be reduced. The proposed model is robust with respect to fault tolerance. Future tasks include system construction and mounting a large-scale integration.

References

1. Mead, C.: Analog VLSI and Neural Systems. Addison Wesley Publishing Company, Inc. (1989)
2. Chong, C.P., Salama, C.A.T., Smith, K.C.: Image-Motion Detection Using Analog VLSI. IEEE Journal of Solid-State Circuits 27(1), 93–96 (1992)
3. Lu, Z., Shi, B.E.: Subpixel Resolution Binocular Visual Tracking Using Analog VLSI Vision Sensors. IEEE Transactions on Circuits and Systems-II: Analog and Digital Signal Processing 47(12), 1468–1475 (2000)
4. Saito, T., Inamura, H.: Analysis of a simple A/D converter with a trapping window. In: IEEE Int. Symp. Circuits Syst., pp. 1293–1305 (2003)
5. Luthon, F., Dragomirescu, D.: A Cellular Analog Network for MRF-Based Video Motion Detection. IEEE Transactions on Circuits and Systems-I: Fundamental Theory and Applications 46(2), 281–293 (1999)
6. Yamada, H., Miyashita, T., Ohtani, M., Yonezu, H.: An Analog MOS Circuit Inspired by an Inner Retina for Producing Signals of Moving Edges. Technical Report of IEICE, NC99-112, pp. 149–155 (2000)
7. Okuda, T., Doki, S., Ishida, M.: Realization of Back Propagation Learning for Pulsed Neural Networks Based on Delta-Sigma Modulation and Its Hardware Implementation. ICICE Transactions J88-D-II-4, 778–788 (2005)
8. Kawaguchi, M., Jimbo, T., Umeno, M.: Motion Detecting Artificial Retina Model by Two-Dimensional Multi-Layered Analog Electronic Circuits. IEICE Transactions E86-A-2, 387–395 (2003)
9. Kawaguchi, M., Jimbo, T., Umeno, M.: Analog VLSI Layout Design of Advanced Image Processing for Artificial Vision Model. In: IEEE International Symposium on Industrial Electronics, ISIE 2005 Proceeding, vol. 3, pp. 1239–1244 (2005)
10. Kawaguchi, M., Jimbo, T., Ishii, N.: Analog VLSI Layout Design and the Circuit Board Manufacturing of Advanced Image Processing for Artificial Vision Model. In: Lovrek, I., Howlett, R.J., Jain, L.C. (eds.) KES 2008, Part II. LNCS (LNAI), vol. 5178, pp. 895–902. Springer, Heidelberg (2008)
11. Kawaguchi, M., Jimbo, T., Umeno, M.: Dynamic Learning of Neural Network by Analog Electronic Circuits. In: Intelligent System Symposium, FAN 2010, S3-4-3 (2010)
12. Kawaguchi, M., Jimbo, T., Ishii, N.: Dynamic Learning of Neural Network by Analog Electronic Circuits. In: König, A., Dengel, A., Hinkelmann, K., Kise, K., Howlett, R.J., Jain, L.C. (eds.) KES 2011, Part IV. LNCS, vol. 6884, pp. 73–79. Springer, Heidelberg (2011)
13. Kawaguchi, M., Jimbo, T., Ishii, N.: Analog Learning Neural Network using Multiple and Sample Hold Circuits. In: IIAI/ACIS International Symposiums on Innovative E-Service and Information Systems, IEIS 2012, pp. 243–246 (2012)
14. Kawaguchi, M., Jimbo, T., Ishii, N.: Analog Real Time Learning Neural Network using Multiple and Sample Hold Circuits. In: Frontiers in Artificial Intelligence and Applications, vol. 243, pp. 1749–1757 (2012)

An Efficient Classification for Single Nucleotide Polymorphism (SNP) Dataset

Nomin Batnyam, Ariundelger Gantulga, and Sejong Oh[*]

Abstract. Recently, a Single Nucleotide Polymorphism (SNP) which is a unit of genetic variations has caught much attention as it is associated with complex diseases. Various machine learning techniques have been applied on SNP data to distinguish human individuals affected with diseases from healthy ones or predict their predisposition. However, due to its data format and enormous feature space SNP analysis is a complicated task. In this research an efficient method is proposed to facilitate the SNP data classification. The aim was to find the most effective way of SNP data analysis by combining various existing techniques. The experiment was conducted on four SNP datasets obtained from the NCBI Gene Expression Omnibus (GEO) website, two of them are from patients with mental disorders and their healthy parents; and the other two are cancer related data. The analysis process consists of three stages: first, reduction of feature space and selection of informative SNPs; next, generation of an artificial feature from the selects SNPs; and last but not least, classification and validation. The proposed approach proved to be effective by distinguishing two groups of individuals with high accuracy, sometimes even reaching 100% preciseness.

Keywords: Single Nucleotide Polymorphism (SNP), classification, feature selection.

1 Introduction

Human genome consists of approximately three billion DNA base pairs, called nucleotide. Nearly 99% of them are identical among all humans (population), and

Nomin Batnyam · Ariundelger Gantulga · Sejong Oh
Department of Nanobiomedical Science and WCU Research Center, Dankook University,
Cheonan 330-714, South Korea
e-mail: {gngrfish,ariuka_family}@yahoo.com
    dkumango@gmail.com

[*] Corresponding author.

R. Lee (Ed.): *Computer and Information Science*, SCI 493, pp. 171–185.
DOI: 10.1007/978-3-319-00804-2_13   © Springer International Publishing Switzerland 2013

only one percent varies among individuals. A large portion of these genetic variations occur as Single Nucleotide Polymorphisms (SNPs). Studies have shown that SNPs may have important biological effects, such as association with complex diseases and different reactions to medications and treatments. Also, it has several advantages over microarray gene expressions, such as it is unlikely to change over time. That is, SNPs of a patient at a birth will remain same whole life. It is much easier and faster to collect SNP sample, for it can be obtained from any tissues in the body, while microarray sample must be taken only from specific tissues [1]. Consequently, gene mapping and detection of polymorphisms have caught much attention recently, and currently enormous number of genetic variations is being discovered and analyzed. Many machine learning techniques have been proposed and applied to SNP data classification. However, it is facing several challenges due to its high volume, which poses computational time complexity and low accuracy.

There is no universally optimal method that fits well with every type of data. Therefore, the aim of this study is to find an efficient way of SNP data analysis, by combining known approaches such as feature selection, R-value evaluation, feature fusion and classification to achieve higher accuracy and less time consumption. In our experiment we utilized two powerful and well-known classifiers k-Nearest Neighbor and Support Vector Machines; and an Artificial Gene Making classifier. However, because the number of samples in a SNP dataset is undue small relative to its attribute size, the curse of dimension occurs. To overcome this problem we reduce a feature space through feature selection. The feature selection plays a crucial role in classification of SNP, since it not only solves the problem of dimension, but also lowers time complexity and facilitates the accuracy improvement by selecting the most informative set of attributes. To perform this task we employed four algorithms, two of which are popular methods, Feature Selection based on Distance Discriminant and ReliefF; and two methods, R-value based Feature Selection and an Algorithm based on Feature Clearness.

The remainder of this paper is organized as follows. In section 2, brief introductions to machine learning techniques that were applied in this study are described. We will provide background information of feature selection and classification algorithms. Datasets and proposed method are described in Section 3. In Section 4 we present experimental results and conclude this work in Section 5.

2   Related Work

2.1   Feature Selection

Like microarray gene expression, SNP data has high dimensionality, but several hundred times bigger, making the data analysis impractical. The whole dataset is composed of informative polymorphisms that are often called Tag SNPs, as well as irrelevant ones. Thus elimination of useless SNPs and extraction of a small subset of discriminative ones will help identify tag SNPs that can be used as bio-markers or other cause associated polymorphisms. This process can be done through feature selection, which also facilitates classification task by reducing a search space.

Feature selection is largely divided into filter and wrapper types. Filter methods are applied on a dataset before classification task and usually evaluate features based on simple statistics such as t- or F-statistics, or p-value [3]. Contrarily, wrapper methods make use of learning algorithm to select a feature subset by incorporating it inside the feature search and selection. The latter approach has an advantage over filter method of showing better performance for particular learning algorithms. However, it is more computationally expensive [4]. In our experiments we used only filter methods. Two popular approaches are: a Feature Selection based on Distance Discriminant (FSDD) [5] and a feature weight based ReliefF [6]; and R-value based Feature Selection (RFS) [7] and an Algorithm based on Feature Clearness (CBFS) [8].

2.1.1   ReliefF

ReliefF algorithms are able to detect conditional dependencies between features. The original Relief [2] algorithm can be used to select nominal and numerical attributes, but it is limited to binary class problems. Meanwhile, the newer and more robust version ReliefF is for multiple class problems and capable of dealing with incomplete and noisy data [6]. Main drawback of Relief algorithms is its time complexity compared to other methods in the literature.

In general, Relief algorithm produces quality estimation for each attribute. To do so, it searches two nearest neighbors for a random sample: one from the same class and one from a different class, and updates quality estimation for each feature depending on the values of the sample and its nearest neighbors.

2.1.2   Feature Selection Based on Distance Discriminant (FSDD)

The main advantage of this algorithm is that it produces as optimal result as exhaustive search methods, in the meantime, it makes use of a feature ranking scheme to solve the problem of computational complexity. The basis of FSDD is to find features with good class separability among different classes as well as make samples in the same class as close to one another as possible. A criterion for feature is a difference between the distance within classes d_w multiplied by a user defined value β, which works as an impact controller and usually set to 2, and the distance between different classes d_b, see Equation (1) [5].

$$d_b - \beta d_w \qquad (1)$$

2.1.3   Feature Selection Based on R-value (RFS)

RFS is a simple feature ranking algorithm based on a dataset evaluation measure R-value [18]. R-value measures the quality of a dataset assuming that the separability of classes is strongly related to category overlap. It captures overlapping areas among classes in a dataset, the lower the value of R the more separable are the classes from one another. Accordingly, RFS algorithm scores the overlapping areas of classes for each feature without considering relationship among features. RFS calculates the R value for a feature F_i by dividing the total number of feature values in a target feature that belongs to overlapping area by total number of samples of given dataset, see Equation (2) [7].

$R(F_i)$ = (total number of feature values in F_i that belongs to overlapping areas)/(total number of samples of given dataset)   (2)

2.1.4   Algorithm Based on Feature Clearness (CBFS)

Likewise the above mentioned algorithms, CBFS measures separability of classes in attributes. It adopts CScore which estimates the degree of correctly clustered samples to the centroid of their class. Its approach works by calculating distance between the target sample and centroid of each class, then compares the class of the nearest centroid with the class of the target samples. Classification accuracy of training data serves as clearness value for a feature [8]. One of CBFS strong points is that it can be combined with other feature ranking algorithms; in our experiment we merged it with the R-value.

2.2   Classifiers

Classification is a class predicting process of a data analysis based on supervised learning method. In order to find a class for unknown sample it uses information of samples that already belong to specific groups. Numerous supervised learning algorithms have been developed, in this study we employed two widely used machine learning techniques, k-Nearest Neighbor (KNN) [9] and Support Vector Machine (SVM) [10]; and an Artificial Gene Making [11] method.

2.2.1   K-Nearest Neighbor (KNN)

KNN is one of the most popular machine learning algorithms and is famous for its simplicity and effectiveness. Also it can be applied in a variety of fields. The algorithm classifies objects based on closest training examples in the feature space and by majority vote of its neighbors. Usually Euclidian distance is used as the distance metric for continuous variables. For a target sample KNN algorithm finds the k closest samples in the learning set and predicts its class by assigning the target sample to the class whose samples are the most common among those k neighbors [3].

2.2.2   Support Vector Machines (SVM)

SVM is a non-probabilistic binary linear classifier based on regularization techniques. To do a classification task, it constructs a hyperplane in a feature space that serves as a boundary between two class samples. The larger the distance between the hyperplane and the nearest sample, the better the classification result. The boundary is determined by the probability distribution or based on the classification of the training patterns. When SVM is applied on a multiclass dataset, the problem can be decomposed into a set of binary problems, and then combined to make a final multiclass prediction [19].

2.2.3   Artificial Gene Making (AGM/Alpha) Method

The Artificial Gene Making method, or sometimes referred to as Alpha, was initially devised for classification of microarray dataset. It is confirmed that a congestion area among classes is one of main reasons of low classification result [18]. The main purpose of Alpha is to reduce this congestion area by adding an artificial gene ($n + 1$) to a dataset with n features, assuming that, for example, a dataset with two dimensions (genes) may become easier to classify if a third dimension is added. First, the dataset is divided into training and test. For construction of a new gene, there is a need to determine two important values α and β, where the first one measures the quality of the original dataset and it expresses a matching ratio between the predicted and original class labels of the training data. The β is dependent on α and estimates the congestion area. The values of a new gene in a training data is calculated by (*original class label*) x β-value, and the gene for test data is calculated by (*predicted class label*) x β-value. After the production of a new gene or feature is done, the classification procedure can be performed [11].

2.3   Feature Fusion Method (FFM)

Feature Fusion (FFM) [20] method is simple, yet proved to be effective in many cases when applied on microarray and other biological datasets. It was confirmed that when attributes or genes are combined with other attributes or genes, they frequently improve the classification accuracy thanks to the mutual information and interaction of features [21]. Therefore, we assume that the same technique could be successfully used on SNP dataset.

The idea of the approach is to produce a new dataset by fusing features of the original one, and the fusion can be done in number of ways. That is, values of one feature can be multiplied, averaged, subtracted, or added up with values of another feature. In addition, the number of features to be fused can range from two to as many as an experimenter ventures to try. For example, in case of two features f_i

and f_j with feature values x_n in Equation (3) a new feature constructed by FFM will be calculated as shown in Equation (4).

$$f_i = \{x_{i1}, x_{i2}, x_{i3}, \ldots, x_{in}\}$$
$$f_j = \{x_{j1}, x_{j2}, x_{j3}, \ldots, x_{jn}\}$$ (3)

FFM (avg): $f_k = \{(x_{i1} + x_{j1})/2, (x_{i2} + x_{j2})/2, (x_{i2} + x_{j2})/2, \ldots, (x_{in} + x_{jn})/2\}$ (4)
FFM $(mult)$: $f_k = \{(x_{i1} \times x_{j1}), (x_{i2} \times x_{j2}), (x_{i2} \times x_{j2}), \ldots, (x_{in} \times x_{jn})\}$

3   Methods and Datasets

3.1   Datasets

The SNP datasets tested in this experiment were downloaded from the NCBI Gene Expression Omnibus (GEO) repository [12]. GEO is a public repository that archives and freely distributes microarray, next-generation sequencing, and other forms of high-throughput functional genomic data.

Our experiment was performed on four Affymetrix Mapping 250K Nsp SNP Arrays: GSE9222 [13], GSE13117 [14], GSE16125 [15], and GSE16619 [16], where the first two are related to mental disorders and the latter two are associated with cancers. Dataset features consist of over 250,000 SNPs and two class labels (case and control). GSE16125's SNP values are in the form of continuous numbers, but the other three arrays have alphabetical format. That is, each sample can have one of four SNP markers: AA, AB, BB, and No Call. AA and BB represent homozygous genotype, the AB represents heterozygous genotype, and No Call is a missing value. Human body contains two copies of each gene, one from father and one from mother. If a mutation occurs in one copy of the gene, it is called heterozygous genotype. However, if both copies of a gene are mutated then that individual is considered homozygous. In order to apply feature selection and classification to these three datasets, the alphabetical format has to be transformed into numerical. There are several ways to do so, but in our case, we adopted a method of three binary values [17] for each genotype shown in Equation (5).

$$AA = 100, \quad B = 010, \quad AB = 001, \quad No\ Call = 000$$ (5)

Here we present two datasets results and other two are included in supplementary material. All four datasets summary can be found in Table 1 and brief description of two datasets which included in this paper is below.

GSE13117 series is composed of 120 cases with unexplained mental retardation along with their healthy parents as control.

GSE16619 series is a breast cancer dataset composed of 105 control and 159 case samples; and approximately 500,000 SNPs.

Table 1 Summary of SNP datasets

No	Dataset name	# of SNPs	# of Samples	Case information	Ref.
1	GSE9222	250,000+	567	Autism (ASD)	[13]
2	GSE13117	250,000+	432	Mental Retardation	[14]
3	GSE16125	250,000+	48	Colon Cancer	[15]
4	GSE16619	500,000+	111	Breast Cancer	[16]

3.2 Methods

Our experiment has three main steps. First of all, a feature selection is performed on the whole data to pick informative set of SNPs and reduce the feature space. Each dataset has undergone a feature selection procedure by algorithms described in previous sections and the top 10 to 100 SNPs were chosen. On the second stage a new dataset is produced from the feature selected data. The details of this step are disclosed below, and the last but not least step is classification. The steps were validated by classification algorithms using 10-fold cross-validation. For KNN we consider the closest 7 neighbors ($k = 7$) because this value was found to produce the best accuracy in most cases. We test SVM with kernel such as linear, polynomial, RBF (radial basis function), and sigmoid. When selecting the linear kernel, we get the best accuracy.

3.2.1 Feature Fusion Method (FFM)

In this study we adopted multiplication (*mult*) and average (*avg*) methods of FFM to experiment on SNP data, i.e. create a new feature from the existing feature pairs by multiplying or averaging feature values for all samples.

The artificial data were produced in two different ways: through Original FFM and R-value FFM. Original FFM means generating new features by multiplying or averaging the combination of the original n features and without performing any additional techniques (Fig. 1). The number of generated SNPs in a new dataset equals $_nC_2.$ The R-value FFM, on the other hand, generates new features only from features that are evaluated and then selected by R-value measures. The latter approach is described in the following subsection.

3.2.2 R-value Evaluation

When generating new attributes from the original ones, the feature space grows dramatically. For example, if we create a combination of feature pairs from a subset of ten ($_{10}C_2$), then the number of newly generated features will be 45, which is over four times bigger than the initial size. Therefore, while producing artificial SNPs, choosing only good ones for further analysis is required. R-value (*Rval*) evaluation is, sort of, performing the role of feature selection during the FFM process. Thus the number of SNPs in the new dataset is random depending on the quality of feature pairs.

All features' R-value is calculated and smaller value indicates a better feature. For example, for features x and y, calculated measure will be $Rval(x)$ and $Rval(y)$ respectively. Then R-value is calculated for every combination of two features constructed by *mult* or *avg* methods $Rval(mult(x, y))$ or $Rval(avg(x, y))$. If the value of a new feature is less than the value of the original single feature, then the new feature combination is taken, otherwise the original is taken. Generic steps for R-value FFM are depicted on Fig. 2.

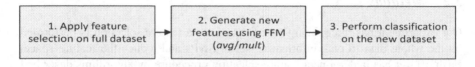

Fig. 1 Generic steps of Original FFM

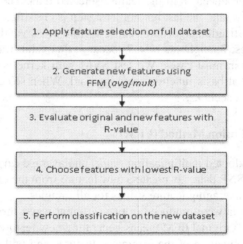

Fig. 2 Generic steps of R-value FFM

4 Results

This section reviews experimental results for datasets: GSE16619 and GSE13117. We will go through datasets one by one. One dataset has three detailed classification accuracy tables - one for each classifier (Supplementary Tables S7 to S18). First column is a feature selection, followed by a FF method, and the last column shows the average accuracy of top 100 SNPs. The top row of accuracy results is Original data, and below are accuracies of new datasets generated by FF methods. The results are compared to the original, and if the accuracy of a new dataset is improved or remained equal, it is marked bold.

4.1 GSE16619 Series

The average classification accuracy for GSE16619 series is summarized in Table 4. From the Table 4 we can see that the average accuracy of the entire new data constructed by FSDD+FFM and classified by Alpha; constructed by RFS + FFM and classified by KNN; constructed by ReliefF + FFM and RFS + FFM and classified by SVM is improved. However, if compare the overall accuracies of FF methods in Table 3 and Fig. 5, then again, the approaches are performing at around similar level, but as for feature selection Table 2 and Fig. 6 show that the CBFS is the best.

Table 2 Comparison of Feature Selection's overall classification accuracy for GSE16619 series. The best results are marked bold.

GSE16619	ReliefF	RFS	FSDD	CBFS
Alpha	0.486667	0.457333	0.486833	**0.665**
KNN	0.51	0.507167	0.565667	**0.684**
SVM	0.580483	0.51285	0.56445	**0.934667**

Table 3 Comparison of Feature Fusion's overall classification accuracy for GSE16619 series. The best improvements are marked bold.

GSE16619	Original	Multiply	Average	Mult+R	Avg+R
Alpha	0.55375	0.482	0.557	0.49325	**0.57425**
KNN	0.57225	0.54775	0.5835	0.55775	**0.59025**
SVM	0.6406	**0.6575**	0.6456	0.64305	0.65075

Table 4 Summary of average classification accuracies of top 100 SNPs for GSE16619 series

GSE16619		Original	Multiply	Average	Mult+R	Avg+R
	ReliefF	0.597	0.411	0.546	0.409	0.545
Alpha	RFS	0.508	0.423	0.47	0.443	0.476
	FSDD	0.473	**0.487**	**0.478**	**0.514**	**0.478**
	CBFS	0.637	0.607	**0.734**	0.607	**0.798**
	ReliefF	0.531	0.481	0.526	0.497	**0.542**
KNN	RFS	0.49	**0.502**	**0.514**	**0.523**	**0.511**
	FSDD	0.561	**0.57**	**0.572**	**0.569**	0.552
	CBFS	0.707	0.638	**0.722**	0.642	**0.756**
	ReliefF	0.5589	**0.5958**	**0.5652**	**0.6057**	**0.5787**
SVM	RFS	0.4941	0.5364	0.5058	0.5058	0.5139
	FSDD	0.5769	0.5535	0.567	0.5598	0.5688
	CBFS	0.9325	**0.9443**	**0.9444**	0.9009	**0.9416**

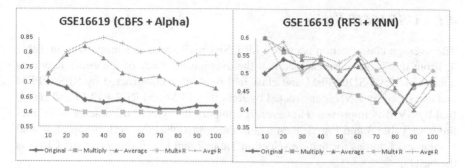

Fig. 3 Accuracy for GSE16619, a) feature selected by CBFS and classified by Alpha (on left); b) feature selected by RFS and classified by KNN (on right) and classified by Alpha. Thick line depicts the original data and other colors are representing new data constructed by FFM, the y and x axes indicate accuracy and number of SNPs respectively.

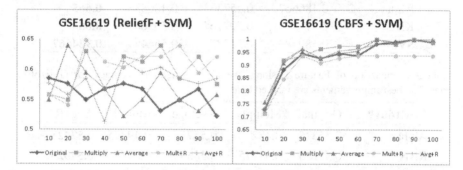

Fig. 4 Accuracy for GSE16619, a) feature selected by ReliefF (on left) and b) CBFS (on right) and classified by SVM. Thick line depicts the original data and other colors are representing new data constructed by FFM, the y and x axes indicate accuracy and number of SNPs respectively.

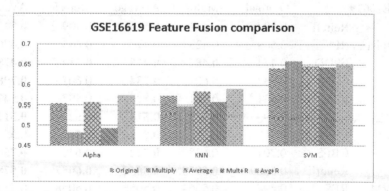
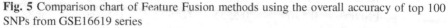

Fig. 5 Comparison chart of Feature Fusion methods using the overall accuracy of top 100 SNPs from GSE16619 series

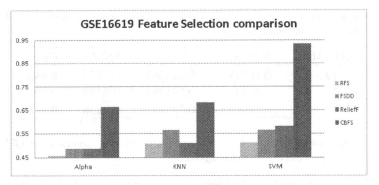

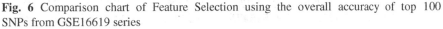

Fig. 6 Comparison chart of Feature Selection using the overall accuracy of top 100 SNPs from GSE16619 series

4.2 GSE13117 series

Due to a high time complexity, we consider only the top 50 original SNPs for generating a new data by R-value FFM. However, from Fig. 7 and Fig. 8 we can see that the said datasets (Mult+R and Avg+R) have a high potential for further accuracy improvement. In addition, Table 5 shows in general that over 4/5 of the newly generated data gave better results than the original.

As for feature selection, CBFS is still performing better than the others, but not outdoing so much compared to other datasets (Fig. 9). Also from Fig. 9 and Fig. 10 we can assume that SVM classifies GSE13117   dataset   the   best,   reaching   99% accuracy (Fig. 8).

Table 5 Summary of average classification accuracies for GSE13117 series

GSE13117		Original	Multiply	Average	Mult+R	Avg+R
	ReliefF	0.6234	**0.6331**	0.614	0.5584	0.549
Alpha	RFS	0.3401	**0.3528**	**0.3497**	**0.353**	**0.352**
	FSDD	0.4555	**0.5288**	**0.4947**	**0.509**	**0.4694**
	CBFS	0.667	**0.6672**	0.671	**0.6674**	0.673
	ReliefF	0.6042	**0.6217**	**0.6045**	**0.6596**	**0.6212**
KNN	RFS	0.5715	**0.5793**	**0.5877**	**0.6282**	**0.602**
	FSDD	0.5863	**0.6382**	**0.6377**	**0.6076**	**0.6048**
	CBFS	0.691	0.6875	**0.6932**	**0.7112**	**0.7286**
	ReliefF	0.6426	**0.9107**	0.6408	**0.86**	**0.6494**
SVM	RFS	0.6404	**0.8664**	0.6396	**0.7346**	**0.6662**
	FSDD	0.7236	**0.9002**	**0.7381**	**0.7832**	0.673
	CBFS	0.8579	**0.9184**	**0.8903**	0.7906	**0.8222**

Table 6 Comparison of Feature Selection's overall classification accuracy for GSE13117 series. The best results are marked bold.

GSE13117	ReliefF	RFS	FSDD	CBFS
Alpha	0.602183	0.349567	0.4977	**0.6688**
KNN	0.62215	0.591383	0.6188	**0.69965**
SVM	0.767633	0.73585	0.7856	**0.8661**

Table 7 Comparison of Feature Fusion's overall classification accuracy for GSE13117 series. The best improvements are marked bold.

GSE13117	Original	Multiply	Average	Mult+R	Avg+R
Alpha	0.5215	**0.545475**	0.53235	0.52195	0.51085
KNN	0.61325	0.631675	0.630775	**0.65165**	0.63915
SVM	0.71612	**0.898925**	0.7272	0.7921	0.7027

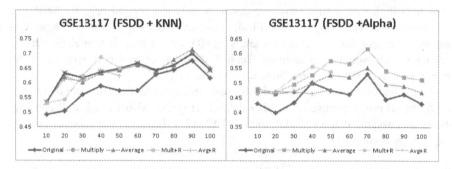

Fig. 7 Accuracy for GSE13117 a) feature selected by FSDD and classified by KNN (on left); b) feature selected by FSDD and classified by Alpha (on right). Thick blue line depicts the original data and other colors are representing new data constructed by FFM, the y and x axes indicate accuracy and number of SNPs respectively.

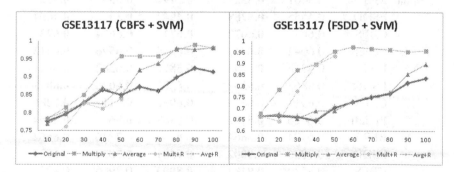

Fig. 8 Accuracy for GSE13117, a) feature selected by CBFS (on left); b) feature selected by FSDD (on right) and classified by SVM. Thick blue line depicts the original data and other colors are representing new data constructed by FFM, the y and x axes indicate accuracy and number of SNPs respectively.

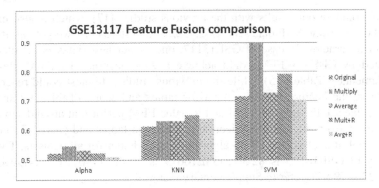

Fig. 9 Comparison chart of Feature Fusion methods using the overall accuracy of top 100 SNPs from GSE13117 series

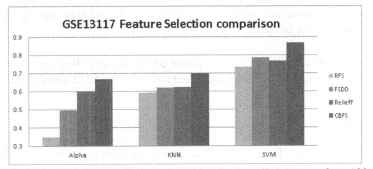

Fig. 10 Comparison chart of Feature Selection using the overall accuracy of top 100 SNPs from GSE13117 series

5   Discussion and Conclusion

Studies have shown that genetic variations are associated with various diseases and they play an important role in the determination of individual's susceptibility to those diseases. SNP is the most common genetic variation, thus machine learning techniques are increasingly applied to identify the interaction between SNPs and complex diseases. However, reaching high classification accuracy for such type of data is not an easy task, since SNP is composed of only four values.

The aim of this research is to suggest an as effective approach as possible to enhance a SNP dataset classification accuracy, and we believe that the goal was attained improving more than the half of the results. The experiment was conducted on four SNP data series that are related to mental disorders and cancers. To predict the diseases we employed classifiers, feature selections, as well as feature evaluation and generation techniques. Some algorithms work well on some data, but may show average performance on others. In general, the approach is composed of three main steps, which are 1) a selection of the most relevant SNPs, 2) a generation of new SNPs from the selected ones, and 3) a classification using a 10-fold cross validation.

If we compare our results with the previous studies [17], which is also experimented on the same SNP data sets (GSE9222 and GSE13117), we got significantly better accuracies. In case of GSE13117, one of our new datasets selected and generated by CBFS + FFM could achieve a 99% accuracy by SVM classifier (Supplementary Table S15), while the previous studies' highest could reach only 66%. From the Feature Fusion comparison tables and charts (Tables 3,7 and Fig. 5,10) we can see the original FFM and R-value FFM perform at around same level, however, from the Feature Selection comparison tables and charts (Tables 2,6 and Fig. 6,9) it is clear that CBFS algorithms is far better for all datasets. The best accuracy of GSE16125 is 82.5%, GSE16619 is 100%, GSE13117 is 99%, and GSE9222 is 78.8% (refer to Supplementary material).

Therefore, from these results we conclude that SNPs can be effectively used to distinguish individuals with complex diseases from the healthy ones, and the proposed approach is efficient for improvement of SNP data classification, especially the new data selected and generated by CBFS + FFM and classified by SVM tend to produce the most favorable accuracies.

Acknowledgments. This work was supported by the National Research Foundation of Korea Grant funded by the Korean Government (NRF-2012S1A2A1A01028576).

References

1. Waddel, M., Page, D., Zhan, F., et al.: Predicting cancer susceptibility from single-nucleotide polymorphism data: a case study in multiple myeloma. Life and Medical Sciences (2005)
2. Kira, K., Rendell, L.A.: The feature selection problem: traditional methods and new algorithm. In: Proceedings of AAAI (1992)
3. Dutoit, S., Fridly, J.: Introduction to classification in microarray experiments. A practical approach to microarray data analysis, pp. 132–149 (2003)
4. Dy, J.G.: Unsupervised feature selection. Computational methods of feature selection, pp. 19–39 (2008)
5. Liang, J., Yang, S., Winstanley, A.: Invariant optimal feature selection: A distance discriminant and feature ranking based solution. Pattern Recognition 41, 1429–1439 (2008)
6. Robnik-Sikonja, M., Kononenko, I.: Theoretical and empirical analysis of ReliefF and RReliefF. Machine Learning 53, 23–69 (2003)
7. Lee, J., Batnyam, N., Oh, S.: RFS: Efficient feature selection method based on R-value. Computers in Biology and Medicine (2012)
8. Seo, M., Oh, S.: CBFS: High performance feature selection algorithm based on feature clearness. PLoS ONE 7(7) (2012)
9. Cover, T., Hart, P.: Nearest Neighbor pattern classification. IEEE 13(1), 21–27 (1967)
10. Chang, C., Lin, C.: LIBSVM – A library for support vector machines (2005), http://www.csie.ntu.edu.tw/cjlin/libsvm/
11. Seo, M., Oh, S.: Derivation of an artificial gene to improve classification accuracy upon gene selection. Computational Biology and Chemistry 36, 1–12 (2011)

12. Barret, T., Edgar, R.: Gene expression omnibus: microarray data storage, submission, retrieval, and analysis. Methods in Enzymology, 352–369 (2006), http://www.ncbi.nlm.nih.gov/geo/
13. Marshall, C.R., et al.: Structural variation of chromosomes in autism spectrum disorder. Am. J. Hum. Genet. 82(2), 477–488 (2008)
14. McMullan, D.J., et al.: Molecular karyotyping of patients with unexplained mental retardation by SNP arrays: a multicenter study. Hum. Mutat. 30(7), 1082–1092 (2009)
15. Reid, J.F., et al.: Integrative approach for prioritizing cancer genes in sporadic colon cancer. Genes Chromosomes Cancer 48(11), 953–962 (2009)
16. Katoda, M., et al.: Identification of novel gene amplifications in breast cancer and coexistence of gene amplification with an activating mutation of PIK3CA. Cancer Research 69(18), 7357–7365 (2009)
17. Evans, D.T.: A SNP microarray analysis pipeline using machine learning techniques. M.S., Computer Science, Ohio University (2010)
18. Oh, S.: A new dataset evaluation method based on category overlap. Computers in Biology and Medicine 41, 115–122 (2011)
19. Mukherjee, S.: Classifying microarray data using support vector machines. A practical approach to microarray data analysis, pp. 166–185 (2003)
20. Batnyam, N., Tay, B., Oh, S.: Boosting classification accuracy using feature fusion. In: 2012 International Conference on Information and Network Technology (ICINT), vol. 37 (2012)
21. Hanczar, B., Zucker, J.D., et al.: Feature construction from synergetic pairs to improve microarray-based classification. Bioinformatics 23, 2866–2872 (2007)

Construction Technique of Large Operational Profiles for Statistical Software Testing

Tomohiko Takagi and Zengo Furukawa

Abstract. This paper shows a novel construction technique of large operational profiles in order to effectively apply statistical software testing to recent projects in which large software is developed in a short timeframe. In this technique, test engineers construct small operational profiles that represent usage characteristics of components of SUT (software under test), and a large operational profile that represents usage characteristics of the whole of SUT is automatically generated from the small operational profiles so as to satisfy behavioral constraints among the components of the SUT. The large operational profile generated by this technique is called a product operational profile, and is used to generate test cases in statistical testing. The key idea of this technique is that an operational profile of the whole of SUT is too large to be constructed manually, but operational profiles of the components of SUT are small and therefore are easy to be constructed. We propose the basic notions of the product operational profile, the behavioral constraints and a generation algorithm, and then evaluate their effectiveness by using an example of software.

Keywords: software testing, model-based testing, operational profile, state machine.

1 Introduction

Failures of software need to be found before shipping in order not to inflict damage on society and also the software development company itself. Software testing [1, 7] is an important technique to find latent failures of software before shipping, and is indispensable to software development. *Statistical testing* (operational profile-based testing) [6, 13] is one of effective and relatively novel software

Tomohiko Takagi · Zengo Furukawa
Faculty of Engineering, Kagawa University
2217-20 Hayashi-cho, Takamatsu-shi, Kagawa 761-0396, Japan
e-mail: {takagi,zengo}@eng.kagawa-u.ac.jp

R. Lee (Ed.): *Computer and Information Science*, SCI 493, pp. 187–199.
DOI: 10.1007/978-3-319-00804-2_14   © Springer International Publishing Switzerland 2013

testing techniques that are classified into model-based testing [8], random testing, and black box testing, and it is used in cases where test engineers need to evaluate and improve software reliability [10]. In statistical testing, test engineers manually construct *operational profiles*, which are probabilistic state machines that represent usage characteristics of SUT (software under test) on expected usage environments. The operational profiles are used to automatically generate test cases as sequences of successive state transitions from an initial state to a final state. The test cases reflect the usage characteristics statistically, and therefore executing those makes it possible to evaluate the reliability of SUT on expected usage environments, and to find failures of SUT that have serious impacts on the reliability. Statistical testing is effectively applicable to integration testing and system testing of SUT that interacts with other systems and/or users and varies its behavior with its states, since state machines are used for modeling the usage characteristics. Also, test engineers can review the design of SUT by analyzing operational profiles. Statistical testing has been applied to real software development, and some literatures show actual results. For example, Popovic et al. [9] and Hartmann et al. [4] applied statistical testing to communication protocols and medical systems respectively, and they achieved the improvement of the reliability of the SUT.

Statistical testing is performed based on operational profiles, and its effectiveness depends on the quality of the operational profiles. Therefore, the construction of operational profiles is one of the most important processes in statistical testing, and it requires great efforts of test engineers that have the expert knowledge of specifications, application domains, and usage environments of SUT. For example, our research group spent over 7 man-days on the construction of an operational profile of commercial software whose scale is about 60 KLOC [12].

In recent years, operational profiles as well as software are becoming rapidly larger. Additionally, recent software tends to be developed in a shorter timeframe, and therefore test engineers cannot avoid reducing their efforts of testing processes. In order to adopt statistical testing and improve software reliability, most software development projects need a technique to construct large operational profiles effectively, but such a technique has not been established.

This paper shows a novel construction technique of large operational profiles in order to effectively apply statistical testing to recent projects in which large software is developed in a short timeframe (hereinafter referred to as *this technique*). In this technique, test engineers construct *elemental operational profiles*, which are small operational profiles that represent usage characteristics of components of SUT, and then a large operational profile that represents usage characteristics of the whole of SUT is automatically generated from the elemental operational profiles so as to satisfy *behavioral constraints* among the components of the SUT. The large operational profile generated by this technique is called a *product operational profile*, and is used to generate test cases in statistical testing.

The key idea of this technique is that an operational profile of the whole of SUT is too large to be constructed manually, but operational profiles of the components of SUT are small and therefore are easy to be constructed. This technique is applicable to not only software that behaves sequentially but concurrent, distributed and asynchronous software that is widely developed and used in recent years.

The rest of this paper is organized as follows. In Sect. 2, we propose the basic notions of product operational profiles, behavioral constraints among components, and a generation algorithm this technique consists of. Sect. 3 illustrates the effectiveness of this technique with an example of software, and Sect. 4 describes related work. Sect. 5 concludes this paper and shows our future work.

2 Technique Overview

This section introduces the basic notions of conventional operational profiles, and then proposes product operational profiles, behavioral constraints among components, and an algorithm to generate product operational profiles.

2.1 Conventional Operational Profiles

An operational profile is a probabilistic state machine that represents usage characteristics of SUT on expected operational environments. The state machine is constructed based on specifications of SUT by test engineers. In the state machine, a node represents a state that characterizes the behavior of the SUT, and an arc represents a transition that is fired by an event. The event corresponds to an input from users and other systems that interact with the SUT, and so on. Also, the probability distributions on the state machine mean the frequency of occurrence of events on each state, and they are determined based on information about application domains, execution logs of other software that is similar to the SUT, and so on.

A simple example of an operational profile is shown in Fig. 1 (b), which is constructed by adding probability distributions to the state machine shown in Fig. 1 (a). In this paper, each state and event is identified by number and alphabet, respectively. Each transition has the occurrence probability of the event in the from-state. For example, the occurrence probability of event b and c in state 1 is 40% and 60% respectively. In the test case generation process, transitions to be executed are randomly selected based on the probability distributions.

2.2 Product Operational Profiles

It is difficult for test engineers to construct large operational profiles to apply statistical testing to large software in a short timeframe. To solve this problem, we propose product operational profiles that are automatically generated from elemental operational profiles so as to satisfy behavioral constraints by using a generation algorithm. The product operational profile is a large probabilistic state machine that represents usage characteristics of the whole of large SUT, and is used to generate test cases for the SUT. The elemental operational profiles are small probabilistic state machines that represent usage characteristics of components of the SUT, and are constructed

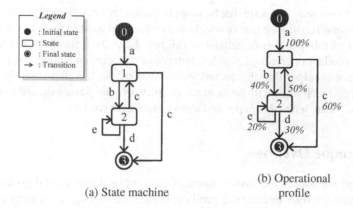

(a) State machine   (b) Operational profile

Fig. 1 Simple example of a conventional operational profile

by using the conventional manner described in Sect. 2.1. The behavioral constraints and the generation algorithm are proposed in Sect. 2.3 and Sect. 2.4, respectively.

The rest of this section shows the basic notions of the product operational profiles. Each elemental operational profile corresponds to each components of SUT, and therefore when there are N ($N > 1$) components to be tested, a set of elemental operational profiles constructed by test engineers is expressed as $EOP = \{eop_1, eop_2, \cdots, eop_N\}$. When a set of states of eop_i ($1 \leq i \leq N$) is expressed as S_i, a set of states of a product operational profile pop is generated by the formula $S' = S_1 \times S_2 \times \cdots \times S_N$. Thus states of pop are N-tuples, and they are expressed as $S' = \{(s_1, s_2, \cdots, s_N) \mid \forall i.1 \leq i \leq N \rightarrow s_i \in S_i\}$. Also, a transition of pop is a 3-tuple and expressed as (s'_f, e', s'_t). s'_f, e' and s'_t mean a from-state, an event and a to-state of the transition respectively, and these have the following properties.

1. $s'_f, s'_t \in S'$.
2. Related behavioral constraints are satisfied.
3. $diff(s'_f, s'_t) = 1$. In this paper, $diff(a,b) = c$ means that there are c different elements between N-tuples a and b. For example, if $a = (1,2,3,4)$ and $b = (1,2,5,4)$, $diff(a,b) = 1$ since the only third element is different between a and b. This property is that one transition of pop always changes one state of one element of EOP.
4. $s'_f[i] \neq s'_t[i] \rightarrow (s'_f[i], e', s'_t[i]) \in T_i$. In this paper, $a[b]$ means the bth element of an N-tuple a. T_i is a set of transitions of eop_i, and its element is similarly expressed as a 3-tuple (a,b,c), where a, b, and c mean a from-state, an event, and a to-state in eop_i, respectively. This property is that each transition of pop always corresponds to a transition of eop_i.
5. If $s'_f[i] \neq s'_t[i]$, the transition probability is given by

$$prob(s'_f, e', s'_t) = \frac{prob(s'_f[i], e', s'_t[i]) \times w_i}{\sum_{x=1}^{N} prob(s'_f, x) \times w_x}, \tag{1}$$

where $prob(a,b,c)$ means the occurrence probability of a transition (a,b,c), and $prob(a,b)$ means the total sum of the probabilities of eop_b's transitions whose from-state is $a[b]$ and that satisfy related behavioral constraints. w_i is the weight of eop_i, that is, the ratio of using the component corresponding to eop_i, and it satisfies the following.

- $0.0 < w_i < 1.0$
- $\sum_{x=1}^{N} w_x = 1.0$

Fig. 2 gives a simple example of elemental operational profiles. SUT represented by this figure consists of two components, and they are modeled as elemental operational profiles eop_1 and eop_2. The weights of eop_1 and eop_2 are $w_1 = 0.4$ and $w_2 = 0.6$, respectively. There is no behavioral constraint among them. The product operational profile generated from eop_1 and eop_2 is shown in Fig. 3. As mentioned above, the state space of a product operational profile is the direct product of all state spaces of elemental operational profiles, and transitions based on the elemental operational profiles connect the generated states. For example, the state marked with (I) in Fig. 3 is expressed as a 2-tuple $(1,4)$, which means that the states of eop_1 and eop_2 are 1 and 4, respectively. eop_1 has the transition $(1,b,2)$, and therefore Fig. 3 has the transition $((1,4),b,(2,4))$, which means that the occurrence of the event b changes the state from $(1,4)$ to $(2,4)$. Transition probabilities of a product operational profile are calculated by the formula (1) included in the property 5. The transition probability of $((1,4),b,(2,4))$ is given by the following.

$$prob((1,4),b,(2,4)) = \frac{prob(1,b,2) \times w_1}{prob(1,b,2) \times w_1 + prob(1,d,3) \times w_1 + prob(4,e,5) \times w_2}$$
$$= \frac{60\% \times 0.4}{60\% \times 0.4 + 40\% \times 0.4 + 100\% \times 0.6} = 24\%$$

The following shows another example of calculating a transition probability.

$$prob((3,5),g,(3,6)) = \frac{prob(5,g,6) \times w_2}{prob(5,f,5) \times w_2 + prob(5,g,6) \times w_2}$$
$$= \frac{70\% \times 0.6}{30\% \times 0.6 + 70\% \times 0.6} = 70\%$$

2.3 Behavioral Constraints among Components

Statistical testing works on the premise that operational profiles are correctly constructed so as to generate only test cases that are executable from the viewpoint of specifications. However, a product operational profile constructed without due consideration about it may generate test cases that are not executable. In general, test cases that are not executable detract from the effectiveness of testing [3, 5]. To solve this problem, we propose behavioral constraints and introduce them to product

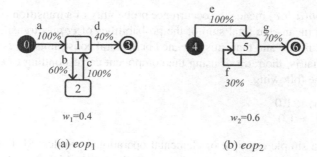

Fig. 2 Simple example of elemental operational profiles

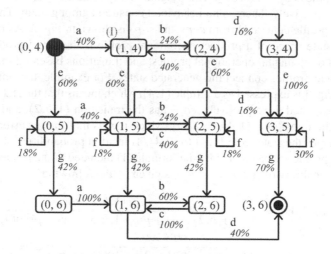

Fig. 3 Simple example of a product operational profile

operational profiles. The behavioral constraints are defined by test engineers, and are used in the generation of a product operational profile and test cases.

The behavioral constraints are classified into the following two sorts.

- A *current constraint* specifies that the occurrence of specific events and/or transitions in an elemental operational profile are permitted/prohibited when current states of other elemental operational profiles are specific ones. This sort of constraint is reflected in the generation of product operational profiles, which is discussed in Sect. 2.4.
- A *historic constraint* specifies that the occurrence of specific events and/or transitions in an elemental operational profile are permitted/prohibited after other elemental operational profiles execute specific events, transitions and/or sequences of transitions. This sort of constraint can hardly be reflected in the generation of product operational profiles, and therefore it is reflected in the generation of test cases. For example, when a generated test case violates a constraint, the test case is replaced with another test case.

```
<constraint>
        ::= <list-of-pre-conditions> "->" <post-condition> ";"
<list-of-pre-conditions> ::= <pre-condition> | <pre-condition>
                        <join-operator> <list-of-pre-conditions>
<pre-condition> ::= <pre-function> |
                    <unary-operator> <pre-function>
<post-condition> ::= <post-function> |
                    <unary-operator> <post-function>
<pre-function> ::= <eop> ".current(" <list-of-states> ")" |
                   <eop> ".historic(" <list-of-objects1> ")"
<post-function> ::= <eop> ".permit(" <list-of-objects2> ")" |
                    <eop> ".prohibit(" <list-of-objects2> ")"
<eop> ::= "eop[" <identification-number> "]"
<list-of-states> ::= <state> | <state> "," <list-of-states>
<list-of-objects1> ::= <object1> |
                       <object1> "," <list-of-objects1>
<object1> ::= <event> | <transition> | <transition-sequence>
<list-of-objects2> ::= <object2> |
                       <object2> "," <list-of-objects2>
<object2> ::= <event> | <transition>
<join-operator> ::= "and" | "or"
<unary-operator> ::= "!" | "forall" <expression> |
                     "exists" <expression>
```

*Note: Some definitions are omitted to save space in this paper.

Fig. 4 Basic structure of the formal language (defined by BNF)

In this study, a formal language based on predicate logic is proposed and is used to define behavioral constraints in order to avoid ambiguity. Also, it is written by using only ASCII characters in order that a computer can easily read and interpret. Fig. 4 shows the basic structure of the formal language we propose, and the following shows some simple examples of behavioral constraints written in the formal and natural languages.

- `eop[1].current(7) -> forall 2<=i<=5 eop[i].permit(b);`
  If the current state of eop_1 is 7, the occurrence of the event b is permitted in all eop_i ($2 \leq i \leq 5$).
- `!eop[2].current(10,11,13) and exists 3<=i<=6`
  `eop[i].current(5) -> eop[1].prohibit((2,f,4),(2,g,5));`
  If the current state of eop_2 is not 10, 11 and 13, and there are one and more eop_i ($3 \leq i \leq 6$) whose current state is 5, then the occurrence of the transition $(2,f,4)$ and $(2,g,5)$ are prohibited in eop_1.
- `eop[1].historic((4,s,6)(6,t,7)) -> eop[2].prohibit(e);`
  After the sequence of transitions $(4,s,6)(6,t,7)$ is executed in eop_1, the occurrence of the event e is prohibited in eop_2.

Fig. 5 shows the product operational profile that is generated from the elemental operational profiles shown in Fig. 2 and the following behavioral constraint.

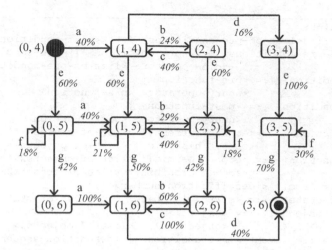

Fig. 5 Simple example of a product operational profile that reflects a behavioral constraint

- `eop[2].current(5) -> eop[1].prohibit(d);`

2.4 Generation Algorithm of Product Operational Profiles

In this section, we propose an algorithm to automatically generate a product operational profile from elemental operational profiles and behavioral constraints. This algorithm consists of the following four steps.

1. Generate the states of a product operational profile. It is given by direct product of all the sets of states of elemental operational profiles. For example, when all the sets of states of elemental operational profiles are $S_1 = \{0,1,2,3\}$ and $S_2 = \{4,5,6\}$, the states of a product operational profile result in $S_1 \times S_2 = \{(0,4),(0,5),(0,6),(1,4),(1,5),(1,6),(2,4),(2,5),(2,6),(3,4),(3,5),(3,6)\}$.
2. Extract all events that can occur in each state generated in Step 1 from elemental operational profiles. Then, if the occurrence of an extracted event does not violate behavioral constraints, add a transition triggered by the extracted event to the product operational profile. For example, all the events that can occur in a state 1 and a state 4 are extracted for the generated state $(1,4)$. Then, if an event b brings the transition from a state 1 to a state 2 and it does not violate constraints, a transition $((1,4),b,(2,4))$ is generated and added to the product operational profile.
3. Calculate transition probabilities by using the formula (1) given in Sect. 2.2 property 5.
4. Detect errors. If there are states that can reach from an initial state and cannot reach to expected final states, or there are nondeterministic transitions, point out them and suggest revising constraints and elemental operational profiles.

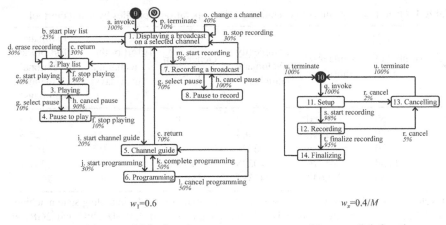

$w_1=0.6$ $w_x=0.4/M$

(a) eop_1 : Main function   (b) eop_x : Sub function

Fig. 6 Elemental operational profiles of video software

3   Case Study

This section illustrates the effectiveness of this technique with an example of software.

First, we introduce video software as the example. The video software contains two sorts of components that behave concurrently. One provides the main function such as manual recording, playing and programming, and another provides the sub function to execute programmed recording in the background. The number of the instance of the former is always one. When the video software is installed on a hardware that includes multiple television tuners, multiple instances of the latter are created and executed. The number of the instances of the latter is expressed as M ($M \geq 1$) in specifications. Therefore, the value of N, that is, the number of components to be tested results in $N = M + 1$.

We constructed elemental operational profiles based on the above, which are shown in Fig. 6. The elemental operational profiles of the former and the latter are referred to as eop_1 and eop_x ($2 \leq x \leq M + 1$), respectively. Also, we defined four behavioral constraints among them. We spent 0.5 man-days on the construction of the elemental operational profiles and the behavioral constraints.

In this study, we have developed a tool to perform statistical testing including this technique. This tool has the functions to automatically generate product operational profiles from elemental operational profiles and behavioral constraints, and generate test cases from them. We generate the product operational profiles of the video software by using this tool, and show its results in Table 1.

As a numerical value given to M becomes larger, the number of states and transitions, the complexity of product operational profiles, and the estimated effort of the conventional manner increase rapidly. The estimated effort of the conventional manner means the man-days that will be spent by test engineers when a large

Table 1 Overview of results of generating product operational profiles by using our tool

M	Number of states	Number of transitions	Complexity (Cyclomatic number) [a]	Generation time (msec) [b]	Effort of this technique (man-days)	Estimated effort of the conventional manner (man-days)
1	50	160	112	35.7	0.5	3.5
2	250	1152	904	83.0	0.5	28.3
3	1250	7526	6278	709.4	0.5	196.2
4	6250	46476	40228	4000.1	0.5	1257.1
5	31250	276642	245394	48956.3	0.5	7668.6

[a] Cyclomatic number represents the structural complexity of graphs including state machines [1]. [b] The execution environment is a laptop computer with 2.5GHz CPU and 2GBytes RAM.

operational profile that corresponds to a product operational profile is manually and directly constructed from the specifications of the video software, and it is calculated based on the following.

- The assumption that the effort spent on the manual construction of an operational profile and the complexity of the operational profile are in proportion
- The effort spent in manually constructing the elemental operational profiles and constraints (The effort can be measured with man-days, and it results in 0.5.)
- The complexity of the elemental operational profiles (The complexity can be measured with Cyclomatic number [1], and it results in 16. Also, we assume that a large operational profile constructed by using the conventional manner has the same complexity as its corresponding product operational profile.)

It is obvious that, even if the value of M is small, an operational profile that corresponds to a product operational profile can hardly be constructed by using the conventional manner, since it requires great effort. On the other hand, this technique generates product operational profiles within one minute, and therefore its effort is always only 0.5 man-days spent on constructing the elemental operational profiles and constraints. Additionally, the automation by this technique would help reduce human errors. It is concluded that this technique enables the construction of large operational profiles in a short timeframe, and the improvement of the applicability and effectiveness of statistical testing.

4   Related Work

Statistical testing [6, 13] is a kind of model-based testing where test cases are systematically generated from formal models that represent the behavior of SUT. Many techniques related to the model-based testing have been proposed and introduced in actual software development [8]. In most of the techniques, test cases are generated based on criteria to cover formal models, such as N-switch coverage to cover all the sequences of successive transitions of length $N + 1$ in a state machine [2]. On the other hand, in statistical testing, test cases are generated and executed based on criteria to reach a specific level of software reliability, such as MTTF (mean time to failure) [14] and the degree of the equality of the probability distributions between an operational profile and testing experience [11]. The criteria to cover formal models do not need to be always used in statistical testing, and therefore, even if a product operational profile is quite large, it does not always result in executing a huge number of test cases. The number of test cases to be executed is determined by the reliability of SUT rather than the size of operational profiles. Also, even if the number of test cases is huge, test case execution can be automated in domain-specific statistical testing [9], and therefore large operational profiles do not impose an additional burden on test engineers.

It is important to give accurate probability distributions to operational profiles, which can be automated by tools. For example, Hartmann et al. [4] gather execution logs of software by using a capture/replay tool that is a kind of test tool, and calculate the probability distributions based on them. The execution logs can also be used to generate accurate operational profiles based on high-order Markov chains [12].

Our technique is compatible with the conventional ones mentioned above, and can be incorporated into conventional statistical testing frameworks.

In this paper, operational profiles are constructed based on state machines, but their representational power may not be enough to represent complex software. Extended state machines are useful to solve this problem, and operational profiles can be constructed based on them. It is known that meta-heuristics is effective to generate executable test cases from the extended state machines [3, 5]. When the extended state machine is introduced into statistical testing, its test case generation algorithm can be constructed based on the meta-heuristics such as genetic algorithms.

5 Conclusion and Future Work

In this paper, we proposed a novel construction technique of large operational profiles in order to effectively apply statistical testing to recent projects in which large software is developed in a short timeframe. This technique consists of basic notions of elemental/product operational profiles, behavioral constraints, and a generation algorithm of product operational profiles. The elemental operational profiles are small probabilistic state machines that represent usage characteristics of components of SUT, and are manually constructed by using the conventional manner. The product operational profile is a large probabilistic state machine that represents usage characteristics of the whole of the SUT, and is used to generate test cases for the SUT. The product operational profile is automatically generated from the elemental operational profiles so as to satisfy the behavioral constraints by using the generation algorithm.

The key idea of this technique is that an operational profile of the whole of SUT is too large to be constructed manually, but operational profiles of the components of SUT (that is, elemental operational profiles) are small and therefore are easy to be constructed. The case study reveals the following.

- Even if the number of components to be tested is small, an operational profile that corresponds to a product operational profile can hardly be constructed by using the conventional manner, since it requires great effort.
- This technique generates product operational profiles within one minute, and therefore test engineers have only to spend their effort on constructing the elemental operational profiles and constraints. The effort is always smaller than the effort of the conventional manner.
- As the number of components (that is, the size of SUT) becomes larger, the differences of the effort between the conventional manner and this technique increase rapidly.

Additionally, the automation by this technique would help reduce human errors. It is concluded that this technique enables the construction of large operational profiles in a short timeframe, and the improvement of the applicability and effectiveness of statistical testing.

In future study, we will extend product operational profiles by introducing extended state machines in order to increase their representational power. Also, we will apply this technique to several software development projects in order to evaluate its effectiveness.

Acknowledgements. This work was supported by JSPS KAKENHI Grant Number 23700038.

References

[1] Beizer, B.: Software Testing Techniques, 2nd edn. Van Nostrand Reinhold (1990)

[2] Chow, T.S.: Testing software design modeled by finite-state machines. IEEE Transactions on Software Engineering SE-4(3), 178–187 (1978)

[3] Doungsa-ard, C., Dahal, K., Hossain, A., Suwannasart, T.: Test data generation from uml state machine diagrams using gas. In: Proc. International Conference on Software Engineering Advances, p. 47 (2007)

[4] Hartmann, H., Bokkerink, J., Ronteltap, V.: How to reduce your test process with 30% – the application of operational profiles at philips medical systems. In: Proc. 17th International Symposium on Software Reliability Engineering. CD-ROM (2006)

[5] Kalaji, A., Hierons, R.M., Swift, S.: Generating feasible transition paths for testing from an extended finite state machine (efsm). In: Proc. International Conference on Software Testing Verification and Validation, pp. 230–239 (2009)

[6] Musa, J.D.: The operational profile. In: Reliability and Maintenance of Complex Systems. NATO ASI Series F: Computer and Systems Sciences, vol. 154, pp. 333–344 (1996)

[7] Myers, G.J.: The Art of Software Testing. John Wiley & Sons (1979)

[8] Neto, A.C.D., Subramanyan, R., Vieira, M., Travassos, G.H.: A survey on model-based testing approaches: A systematic review. In: Proc. 1st ACM International Workshop on Empirical Assessment of Software Engineering Languages and Technologies, pp. 31–36 (2007)

[9] Popovic, M., Basicevic, I., Velikic, I., Tatic, J.: A model-based statistical usage testing of communication protocols. In: Proc. 13th Annual IEEE International Symposium and Workshop on Engineering of Computer Based Systems, pp. 377–386 (2006)

[10] Rook, P.: Software Reliability Handbook. Elsevier Science (1990)

[11] Sayre, K., Poore, J.H.: Stopping criteria for statistical testing. Information and Software Technology 42(12), 851–857 (2000)

[12] Takagi, T., Furukawa, Z.: Construction method of a high-order markov chain usage model. In: Proc. 14th Asia-Pacific Software Engineering Conference, pp. 120–126 (2007)

[13] Walton, G.H., Poore, J.H., Trammell, C.J.: Statistical testing of software based on a usage model. Software Practice and Experience 25(1), 97–108 (1995)

[14] Whittaker, J.A., Thomason, M.G.: A markov chain model for statistical software testing. IEEE Transactions on Software Engineering 20(10), 812–824 (1994)

Intelligent Web Based on Mathematic Theory

Case Study: Service Composition Validation via Distributed Compiler and Graph Theory

Ahmad Karawash, Hamid Mcheick, and Mohamed Dbouk

Abstract. This paper discusses a model for verifying service composition by building a distributed semi-compiler of service process. In this talk, we introduce a technique that solves the service composition problems such as infinite loops,deadlock and replicate use of the service. Specifically, the client needs to build a composite service by invoking other services but without knowing the exact design of these loosely coupled services. The proposed Distributed Global Service Compiler, by this article, results dynamically from the business process of each service. As a normal compiler cannot detect loops, we apply a graph theory algorithm, a Depth First Search, on the deduced result taken from business process files.

Keywords: SOA (Service Oriented Architecture), Compiler, Business Process Execution Language (BPEL), Depth First Search (DFS), Distributed Global Service Compiler (DGSC).

1 Introduction

Web services are defined as self-contained, modular units of application logic which provide business functionality to other applications via an Internet connection. Web services support the interaction of business partners and their processes

Ahmad Karawash · Hamid Mcheick
Department of Computer Science, University of Quebec at Chicoutimi (UQAC),
555 Boulevard de l'Université Chicoutimi, G7H2B1, Canada
e-mail: {ahmad.karawash1,hamid_mcheick}@uqac.ca

Mohamed Dbouk
Department of Computer Science, Faculty of Sciences (I), Lebanese University,
Rafic-Hariri Campus, Hadath-Beirut, Lebanon
e-mail: mdbouk@ul.edu.lb

R. Lee (Ed.): *Computer and Information Science*, SCI 493, pp. 201–213.
DOI: 10.1007/978-3-319-00804-2_15   © Springer International Publishing Switzerland 2013

by providing a stateless model of atomic synchronous or asynchronous message exchanges (B. Srivastava and J. Koehler). Every service consists of a domain of computers and computers in a domain can just communicate with each other through predefined functions (Amirjavid F. et al., 2011). Services can be invoked by other services or applications. Services are designed for interaction in a loosely coupled environment, and therefore are an ideal choice for companies seeking inter or intra business interactions that span heterogeneous platforms and systems (K. LI, 2005).

In many cases, a single service is not sufficient to the user's request and often services should be combined through service composition to achieve a specific goal. Nowadays, researches show that problems of composing web services are expressed in designing, discovery, validation and optimization of service composition. The paper proposes a way of dynamic validation of service composition to prevent some errors (infinite loops, blocked services, incorrect business flow design and many others) at the beginning of the design phase of new composite service.

Before designing a new composite service, the service discovery process returns a set of candidate services and at some of those services are used in the composition process according to non-functional criteria (cost, time, and context-aware). But nothing, in this discovery notifies service developer if the invoked services are working normally or not.

In the dynamic world of service-oriented architectures, however, what is sure at design time, unluckily, may not be true at run time. The actual services, to which the workflow is bound may change dynamically perhaps in an unexpected way, may cause the implemented composition to deviate from the assumptions made at design time. Traditional approaches, which limit validation to being a design time activity, are no longer valid in this dynamic setting. Besides performing design-time validation, it is also necessary to perform continuous run-time validation to ensure that the required properties are maintained by the operating system. The compiler is the only way to validate the sequence of service process. It is the program that translates one language to another. Thus our goal is to implement a distributed dynamic compiler that compiles the composition of every new composite service. When a client designs a new composite service, the related compiler Grammar rules, of the invoked services, are sent to him as XML files then combined together to constitutes a local compiler that validate new service composition at design phase.

Section 2 describes the previous methods of service validation. Section 3 gives two service composition examples to highlight the problem of service validation. While section 4 proposes a new service composition validation model in the service design development phase; two techniques of parsing and depth first search are described and a simulation steps are given in this section. Section 5 summarizes the ideas as a conclusion and gives perspectives for future works.

2   Background

This part describes the previous works and some basic features.

2.1   Related Works

Some ways of dynamic system validation are discussed in this section. In 2006 Colombo et al., undertake the topic of dynamic composition where the service parts do not always behave along expected lines. They provide an extension to the BPEL language in the form of the 'SCENE platform' which addresses this issue. The proposed platform is validated forming an application using a set of real services and observing the behavior of the application (Colombo et al, 2006). In 2009, Silva et al proposed the DynamiCos structure which response the requirements of different customers to dynamically put together personalized services. To confirm the proposed structure they set together an extensive model of the structure which enables services to be deployed and be published in a UDDI-like registry (Silva et al, 2009).

In 2008, Eid et al explain a set of scales alongside which to evaluate the various frameworks of dynamic composition. The set of scales is inclusive and is roughly classified into three parts: input subsystem, composition subsystem, and execution subsystem. To be considered good a composition model must achieve well against these scales (Eid *et al.*, 2008).

In 2007, Shen et al found the Role and Coordinator (WSRC) model to hold dynamism in web service compositions. In this model, the development of service composition is divided into three layers: Service, Role, and Coordinator. To validate the model, the authors describe a case-study of a vehicle navigation system which comprises a global positioning system and a traffic control service (Shen *et al.*, 2007).

These are a small list of the validation methods in use for dynamic composition models and structures. Some of these methods are quite complex like the model proposed by DynamiCos or that of the SCENE platform. While, the validation ways that are simple like the case-study in the WSRC model seem useless.

The validation model proposed in this paper, Global compiler service, relatively simple as it expands the validation in one operating system to achieve a wide convention between all the operating systems that deal with service composition.

2.2   Basic Features

Business process execution language (BPEL) - BPEL is a language created to compose, orchestrate and coordinate web services. It is the result of over ten years of collaborative effort in Business Project Management by Microsoft and IBM. It provides both synchronous and asynchronous interactions and it gives suitable forms to create a process. It allows the creation of composite processes with all its related activities. BPMN steps: invoking web services, waiting for clients to invoke the service via message, generate response, manipulate variables, throws exceptions, pause for a selected time, terminate.

Compiler - is a program that takes a source program typically written in a high-level language and produces an equivalent target program in assembly or machine language (Aho I. et al, 2007). Also it is reports error messages as a part of the translation process. A compiler performs two major tasks: analysis of the source program and synthesis of the target-language instructions. In order to build a compiler, there are six phases to follow as in figure 1: i) scanning the input program will be grouped into tokens, ii) parsing or syntax analysis, iii) building a Context-Free-Grammar, iv) applying semantic analysis to keep on mapping between each identifier of data structure (symbol table) and all its information and ensure consistent, v) extracting assembly code generation and vi) finally realizing code optimization.

Depth First Traversal (DFS) – it is a graph theory algorithm for traversing a graph. It is a generalization of preorder traversal. It starts from a vertex and recursively it build a spanning forest that determine if the graph is cyclic (contain cycle loop) or acyclic.

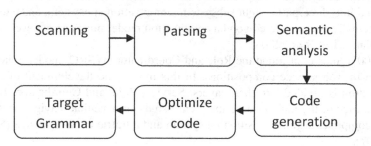

Fig. 1 Phases to build a compiler

3   Composition of Web Service

In order to highlight on the problem of web service composition and simplify the idea for the reader, this section gives two examples about service composition. The first example reflects a simple normal composition while the second shows an abnormal service behavior.

3.1   *Simple Services Composition Example*

Figure 2 shows a simple example of how Providers of web services are communicated to achieve a composed service.

Let $Client_2$ has to solve two mathematical formulas: *"F1: A = 2*x +3*y"* & *"F2: B = 2*x "*

In order to achieve his goal $Client_2$ will design a new composite web service. First of all, he searches in $UDDI_2$ which gives him a summary about the services that are existed in the $Provider_2$. $UDDI_2$ has two services that solve two equations: EQ_1:*"2*x"* & EQ_2: *"3*y"*

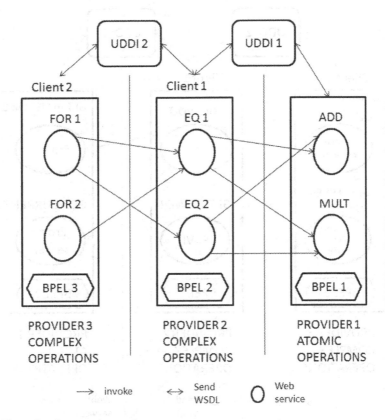

Fig. 2 Example of composite services

Using the information given in the *WSDL* file by $UDDI_2$, $Client_2$ invokes $Provider_2$ operations. But the two services EQ_1 and EQ_2 invoke other services *ADD* & *Multiply* from the $Provider_1$ to complete the required answer. This is a simple idea about how service composition works.

3.2 Infinite Loop Example

Web services are distributed through the whole internet and controlled by various sides. In the modern state, services are dynamically managed. Because the most used services are big and composite, the states of failure and infinite loop are detected sometimes. Failure of composite services results from an obstacle in one of its parts, while infinite loop exists as a result of wrong process flow design.

A real example of the service composition problem (infinite loop) is the TIBCO web service. See the link below (https://www.tibcommunity.com/ message/70086). Figure 3 shows an infinite loop (or cycle) while executing composite service.

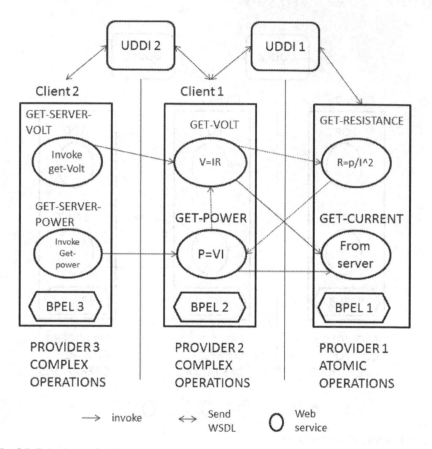

Fig. 3 Infinite loop of web service

Let voltage represented by V, current by I, resistance by R and represent power by P.

We have a set of service to use:

-     $Client_2$ build two services **GET-SERVER-VOLT** & **GET-SERVER-POWER**
-     $Provider_2$ provides two services **Get-Volt** ($V= I*R$) & **Get-Power** ($P=V*I$)
-     $Provider_1$ provides two services **Get-Resistance** & **Get-Current** ($R=p/I\wedge2$)

$Client_2$ wants to calculate the consumption of **Voltage and Power** of the last service provider machine during a composite service process. To complete the needed service, *Client₂* invokes services from *Provider₂* while *Provider₂* invokes other services of *Providerᵢ* to answer the question of *Client₂*. To build his own services (***GET-SERVER-VOLT*** & ***GET-SERVER-POWER***), *Client₂* firstly searches in UDDI₂ about services and invokes **Get-Volt** & **Get-Power** from *Provider₂*.

Regarding the service *"Get-Volt"*, it invokes *Providerᵢ* (the last service provider in this process) services specifically the *"Get-Resistance"* service to calculate resistance '*R*' and it invokes the "**Get-Current**" service to calculate current '*I*'.

From the other side, the service *"Get-Resistance"* invokes *"Get-Power"* service from *Provider₂* in order to calculate power '*P*'. But the service *"Get-Power"* invokes *"Get-Volt"* service to calculate voltage V.

Indeed, the *"Get-Server-Volt"* service falls into an ***infinite loop*** as seen above (figure 3) in red color. The *"Get-Volt"* node invokes the *"Get-Resistance"* node which needs results from *"Get-Volt"*. Thus *"Get-Volt"* invokes itself indirectly. There are also other types of errors may occur because of partial fail or bad service communications.

4   Distributed Global Service Compiler (DGSC)

Our DGSC model consists of extracting compiler Context-Free-Grammar rules of the business process (BPEL) of a web service (figure 4). Then save these rules in the UDDI registry. Grammar rules are used later by the client when he fetches the registry to build a new composite service.

DGSC is simply a verification of new composite service before the execution that is valuable. Beside the development of programming languages, there exist many tools today, for example ATLAS in Eclipse, that transforms the design phase of service to related code automatically. But how achieve the same for on-line dynamic service composition without dangerous errors? As we know, it is impossible to discover the business process for web service through SOA model (meta-data about service process is just published by WSDL file).

Our goal is to discover design errors in the design phase of composite service without knowing the exact flow of service process. In fact, there are many obstacles facing our DGSC because the service design takes place in the client side and the content of web services is dynamically edited from several sides.

Service designer (Client side) searches the UDDI in order to build a new composite service. But nothing verifies that the new combination of service, that may also invoke other services, is free of errors and infinite loops. Also even if a correct composition of a complex service is achieved, this action may fail later because services are dynamically edited.

The proposed solution uses two phases of compiler design (scanning and parsing). This phase of the compiler is applied in the business process (BPEL) of service that contains the internal service design. A grammar rules drive similar to the case of the third phase of compiler design (Context-Free-Grammar). These rules are sent to UDDI registry in XML format. But the client uses the WSDL files of several services to design a new composite service. Thus depending on our

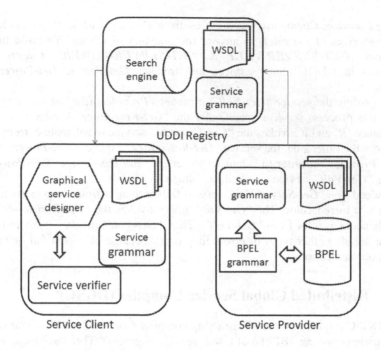

Fig. 4 Distributed global Service Compiler (DGSC)

proposed model he downloads the rules file, from the UDDI, with the WSDL file
and he uses these rules to compile a new design of composite service. Locally on
the client side, a mathematical algorithm (DFS) is applied in these rules to detect
if the new design of composite service contains infinite loop before service
deployment.

4.1 Extraction of Service Grammar

For every programming language there exists a compiler Grammar that is used to
verify the steps of building a new program. As mentioned in section 2.2, there are
six phases to build a compiler. But in our distributed compiler model we just need
to detect infinite loops. Thus scanning then parsing is applied to the BPEL file and
a result is a context-Free-Grammar for the service (similar to the third phase of
building compiler).

To achieve our BPEL parser, we used the BPEL grammar of BPLE4WS written
by the members of the ReDCAFD Laboratory at the University of Sfax.

The BPEL parser is implemented using Netbeans 6.9.1 and Java code. The outcome of the parser is a database entry in a specific table.

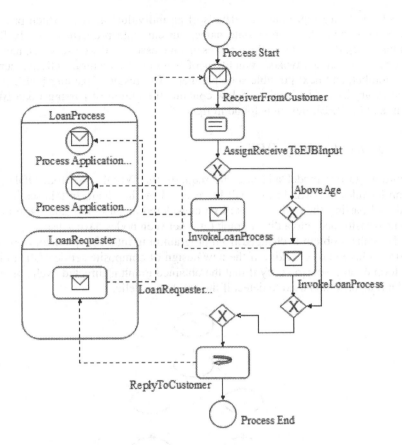

Fig. 5 Loan BPEL example

Table 1 The Output result of parsing the Loan BPEL code

Activity name	0	1	2	3	4	5
Activity Name	receive	if	invoke	if	invoke	Reply
Current State	ReceiveFrom-Customer	nul l	invokeloan-Process	nul l	invokeloan-Process	ReplyToCus-tomer
Conditions	LoanRequesto-rOpera.	nul l	ProcessAppli-cation	nul l	ProcessAppli-cation	LoanProces-sor
Partner Link	LoadRequestor	nul l	LoanProcessor	nul l	LoanProcessor	LoanProces-sor
Opera-tions	LoanRequesto-rOpera.	nul l	ProcessAppli-cation	nul l	ProcessAppli-cation	LoanProces-sor
Next Activity	1	2,3	5	4,5	5	6

Each row entry represents the details of an individual activity which provides information about the current state name, current state properties (as My Role, Partner Role), PartnerLink (which represents the associated web service), name of the operation being invoked, condition of a looping structural activity, current state number, and next possible state numbers. The result of parsing BPEL file is saved in an Excel file. Table 6 below contains the output of parsing Loan BPEL file in and the BPEL design is found in figure 5.

4.2 Detect Cycle by DFS

According to our model and after he requests the WSDL file from UDDI, the Compiler rules in XML format will be received by the client. In this section, the results of parsing the BPEL file will be considered as an inputs and these inputs will be transformed into a direct graph (arcs between nodes have sense).

Now the problem is changed from programming into a graph theory problem (figure 6).Instead of checking if the new design of composite service falls in infinite loop or not, we can verify if that the obtained graph is directed cyclic or acyclic. DFS algorithm is used to detect if the graph is acyclic.

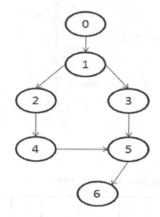

Fig. 6 The directed graph of the BPEL example

Indeed, DFS starts from the root node and explores siblings as far as possible along each branch before backtracking and if it arrive a visited node again it will notify that the graph is cyclic (it contains cycle). But sometimes the service designer need to have a cycle like while-loops, for-loops or even reply to node that sends a request. Thus in all cases we give the designer the permission to discard the detected harmful loops.

4.3 Distributed Compiler Concept

Composite services are built by invoking other already implemented services. But the web services are dynamic and able to be edited at random time.

In our case, the compiler is decentralized (federated) and the Grammar of every service works as a small compiler. When service designer forms a new composite service and the Grammar rules of all what is needed to be invoked services are collected at the client's machine. These rules are combined and DFS is applied to detect errors (infinite loops, errors...etc.). If the result returns an error then a notification appears to the web service designer that he must change the wrong graphical service design and the compiler gives him details about the error.

DGSC deals with the existing implemented services as standards, which have correct design, for new composite service.

In other words, if a developer wants to develop new composite service called *XY* and he needs to use other services then all these other services will stay non-editable at the last stage of designing this service. Thus the developer will receive an update message if the service he needs to invoke change during his design of new composite service. There several scenarios may occur during the design phase of new composite service:

- The relations between the services to be invoked are edited.
- One of the services to be invoked is being deleted or failed.
- An internal change occurs in the behavior of one of the services to be invoked.

Thus the server sends updates to a web service designer about any change occur in the required services of the new composition. Also the server prevents changes in these needed services while the deployment phase of the new composite service takes place. After the deployment of this new service it will be standard for other new services during the design phase.

4.4   Steps of DGSC Simulation

In this simulation we have used Netbeans6.9.1, Apache Tomcat server, Apache-ode-war-1.3.5. We use http://ode.apache.org to execute a BPEL file based on the Apache server.

There are several steps in our simulation (check the sections 4.1 for detail of this simulation):

1. Start the Apache server and Netbeans.
2. Create database then a table (we used EXCEL file) to save the result of parsing a BPEL.
3. Apply the database connection between Netbeans and the database.
4. Use existing or Design some BPEL processes examples (as LoanProcess in our case) in service composition
5. Execute the code of our parser and put the BPEL file which previously designed as input and choose the excel database to save the output.
6. The XWTransformer class transforms the output result in the table to XML format and inserts its content to the WSDL file of the service.

7.   Apply SOAP to send the WSDL content file from server to client (for more details refer to http://ode.apache.org/war-deployment.html).
8.   Execute WXExctractor class code that extracts the parsing result from the WSDL file.
9.   Combine the parsing results of all the services invoked in this composition.
10. Execute the Depth first search (DFS) code using the combined result as input.

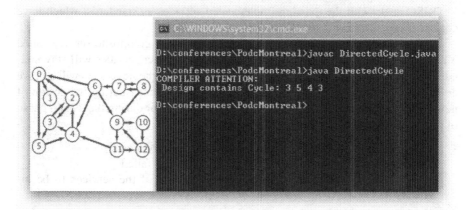

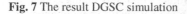

Fig. 7 The result DGSC simulation

This simulation validates the composition of new services. Before deploying the new design composite service, if the simulation detects infinite loop, it will notice the developer to fix it and then change the design. This model helps designer to avoid the cyclic composition service in the real world where a huge number of invocations take place and all the design services are dynamically changed and composed. Figure 7 shows the result of the DGSC simulation that is applied on a group of BPEL files of the services to be invoked to form a new composed service.

5   Conclusion and Future Works

The web service revolution simplified the building of new complex services because it gives the facility to invoke any type of services. On the other side, this revolution offers only the meta-data to the user about the service to be invoked. Thus invoking blindly another service may not give always the needed result. From this point, we start searching about a way to validate the web service at the design phase and the answer is the Distributed global Service Compiler. Since we have millions of services we cannot build a centralized compiler for all the services, we have developed a dynamic compiler per each new composite service. To deal with business process of the service, we started from BPEL and extract the Grammar rules, which represent the internal map (sequence) of that service. As the client contacts the Registry of services to get summary (such as WSDL file) before building a service, DGSC proposes to insert the Grammar rules in the WSDL files of the registry. These rules will be downloaded later by the user (when he invokes WSDL file) at the beginning of the design phase. The DGSC model helps us to validate sequence of the services.

As a future work, our goal is to achieve an intelligent organized web depending on the theorems of Mathematic. Taking this article as a case study, we reach a logical composition of service that can change its job according to client response without destroy the behavior of other services.

References

Srivastava, B., Koehler, J.: Web Service Composition - Current Solutions and Open Problems. In: IBM India & Switzerland Research, ICAPS 2003 Workshop on Planning for Web Services, vol. 35 (2003)

Amirjavid, F., Mcheick, H., Dbouk, M.: Job division in service oriented computing based on time aspect. Int. J. Communication Networks and Distributed Systems 6(1) (2011)

Li, K.: LUMINA: Using WSDL-S For Web Service Discovery. Master Thesis; University of Georgia (December 2005)

Colombo, M., Di Nitto, E., Mauri, M.: SCENE: A service composition execution environment supporting dynamic changes disciplined through rules. In: Dan, A., Lamersdorf, W. (eds.) ICSOC 2006. LNCS, vol. 4294, pp. 191–202. Springer, Heidelberg (2006)

Silva, E., Pires, L.F., van Sinderen, M.: Supporting dynamic service composition at runtime based on end-user requirements. Centre for Telematics and Information Technology University of Twente, The Netherlands P.O. Box 217, 7500 AE Enschede (2009)

Eid, M.A., Alamri, A., El-Saddik, A.: A reference model for dynamic web service composition systems. International Journal of Web and Grid Services (2008)

Shen, L., Li, L., Ren, S., Mu, Y.: Dynamic composition of web service based on coordination model. In: The Joint International Conferences on Asia-Pacific, Web Conference and Web-Age Information Management (2007)

Aho, A.V., et al.: Compilers, principles, techniques, and tools, QA76.76.C65A37 (2007)

Straightening 3-D Surface Scans of Curved Natural History Specimens for Taxonomic Research

James Church, Ray Schmidt, Henry Bart Jr., Xin Dang, and Yixin Chen

Abstract. Two challenges for taxonomists are proper identification of specimens to known species and extracting information from specimens to diagnose new species. Both tasks are complicated by the very large numbers of known and unknown species and the dwindling numbers of qualified taxonomists to identify/diagnose them all. Automated species identification is a tool that can assist taxonomists facing this challenge. This paper looks at one aspect of automated species identification: unfolding curved specimens, which commonly occurs when specimens are prepared for storage in natural history collections. Here we attempt to address the rather extreme case of an elongate fish specimen coiled along its medial axis. The medial axis is the set of all points within an object with the shortest distance to at least two different points on that object's surface, where "distance" (typically Euclidean) is determined by the application. Medial Axis Estimation is a challenging problem that arises when the surface itself is sampled (i.e. incomplete). In this paper, we look at various techniques for estimating the medial axis of an object, then we propose a new method for medial axis estimation based on localized spatial depth. We extend the idea of localized spatial depth-based medial axis further by applying an original ridge detector. We conclude with a comparison of our approach with The Power Crust approach using artificial data.

Keywords: medial axis, ridge detection, taxonomic research.

James Church · Yixin Chen
Department of Computer and Information Science, University of Mississippi, University, MS 38677, USA
e-mail: jcchurch@go.olemiss.edu, ychen@cs.olemiss.edu

Ray Schmidt · Henry Bart Jr.
Department of Ecology and Evolutionary Biology, Tulane University, New Orleans, LA 70118, Tulane University Biodiversity Research Institute, Belle Chasse, LA 70037, USA
e-mail: {rschmidt,hbartjr}@tulane.edu

Xin Dang
Department of Mathematics, University of Mississippi, University, MS 38677, USA
e-mail: xdang@olemiss.edu

R. Lee (Ed.): Computer and Information Science, SCI 493, pp. 215–229.
DOI: 10.1007/978-3-319-00804-2_16   © Springer International Publishing Switzerland 2013

1 Introduction

Taxonomy is a field of biological study in which specimens are classified in groups based on unique characteristics that members of each group share in common and unique names are assigned to identify each group. The field of taxonomy is confronted with several challenges [1] [2]. First, many parts of the world are unexplored by taxonomists and some of these areas are rich with undiscovered species. Second, the pace of taxonomic discovery, as traditionally practiced, has been slow, and the number of practicing taxonomists has been in decline for several decades, resulting in what has been termed the *taxonomic impediment*. Third, human destruction of natural habitats, especially in species rich areas, has resulted in a *biodiversity crisis*, and it is feared that many species will go extinct before they can be discovered and described. Computer tools have the potential to assist taxonomists by automating and expediting the process of diagnosing specimens as members of either known or unknown species [3] [4]. In one computer-aided approach, a specimen is scanned using a 3D scanner, the scan is digitally landmarked, geometric features are extracted from the landmarks and analyzed using a heuristic function. If the heuristic function fails to identify the specimen as a member of a known species with a certain degree of confidence, it is possible that the specimen is representative of a new species. The specimens used in taxonomic studies are typically preserved specimens obtained from natural history museums. Often natural history specimens are preserved with different degrees of curvature of their bodies. Elongate specimens such as eels preserved in jars often take the highly curved shape of their containers, making it difficult to extract features from 3-D scans of these specimens. This paper addresses the challenge of straightening curved natural history specimens. Our goal is to estimate the medial axis of a 3-D scan of an elongated fish specimen for the purposes of unfolding the specimen.

The medial axis is the set of all points within an object with the shortest distance to at least two different points on that object's surface, where "distance" (typically Euclidean) is determined by the application. It is an important tool in computer vision applications in order to determine the "skeleton" of a shape or to approximate surface reconstruction. Applications of the medial axis range from object unfolding to automatic rigid skeleton formation intended for physics applications. In this paper, we focus on medial axis estimation with the given assumption that our input shape consists of only sample points along the surface and that holes exist. The surface scans are often created using 3D scanners and the problems related to scanners are discussed by Bajaj et al. [5].

1.1 Medial Axis Estimators

The medial axis was first proposed by Blum [6] using an idea that he called the "grass fire" analogy (now called the "prarie fire" analogy): if a fire is lit along the perimeter of a shape, it should burn inward, leaving the internal skeleton of the object at the points where the fire is quenched. Boissonnat [7] first proposed that the medial axis of a shape can be determined using the set of points produced by a Voronoi diagram that also lie on the interior of that shape. Edelsbrunner and Mücke [8] took Boissonnat's ideas and gave them a more formal definition using the α-shape algorithm. Each of these approaches assume that a complete knowledge of the shape exists.

Often a perfect knowledge of an object's surface does not exist. The scanning process of a object is sampled from different vantage points and noise is introduced, leaving the object incomplete. For this reason, medial axis estimation techniques were developed. The relevant literature focuses on Delaunay or Voronoi decomposition of points in order to reconstruct the medial axis and surface reconstruction. Amenta et al. [9] created The Power Crust which uses the Voronoi decomposition and focuses on narrow structures in shapes that lead to confusion as to which points are "interior" or "exterior". The Tight Cocone algorithm developed by Dey and Goswami [10] uses the Delaunay triangulation in three dimensions to create a convex hull of the shape. Then the algorithm removes tetrahedrons that lie on the exterior of the shape, leaving an approximation of the surface. The λ-medial axis by Chazal and Lieutier [11] is an attempt at approximating the medial axis of a noisy set of surface points by first sampling the points and then finding the internal set of Delaunay points.

1.2 Ridge Detectors

Ridge detectors are an element of computer vision that help to simplify the analysis of images so that ridges may be applied to unique applications. Most importantly, we focus on the work of Lindeberg [12] describing edge and ridge detection in two dimensions. While there is some work on extending the work of Lindeberg in three dimensions [13], it appears that previous attempts at this use ad hoc methods for formulating a 3D ridge. What we present is a complete work on extending ridge detection to three dimensional images and we set ourselves up for future work in ridge detection beyond three dimensions.

Ridge and edge detection are closely related. Canny [14] sets out the two criteria for an edge in two dimensions. First, the edge detector should be robust with respect to noise. Second, the edge detector should be accurate in order to localize each edge. We apply the same criteria to a ridge detector in three dimensions.

Traditionally a ridge is defined in two dimensions as a raised separator (e.g. a mountain range) which divides two distinct regions. The ridge itself extends in two directions orthogonal to the two regions. Describing the property of a ridge in three dimensions requires us to think abstractly. We define a ridge as a raised separator which divides an area along two directions which are orthogonal to each other. A third direction orthogonal to both of the two directions represents the direction of the ridge itself. In an n-dimensional setting, we define a ridge as a raised separator which divides an area along n-1 orthogonal directions. The nth-direction (also orthogonal) is the direction of the ridge. Based on this definition, we can set the following criteria that must be met for a ridge: (1) There should be a negative gradient change in two orthogonal directions from a point; (2) In the direction orthogonal to this gradient change, there should be no significant gradient change; (3) This should be invariant to scale; (4) This should be robust with respect to noise.

1.3 Paper Organization

The remainder of this work is organized as follows. In section 2 we introduce the rational for using the localized spatial depth rather than the Voronoi decomposition. In section 3 we provide a formal definition for a ridge in three dimensions. In section 4 we outline an approach to unify chains of ridges. In section 5 we provide a complete algorithm for our approach to medial axis estimation and ridge alignment. In section 6 we provide a comparison of our approach with artificial data sets and compare them to ground truth as well as show examples our approach on using real data. In section 7 we discuss potential open questions introduced by our research.

2   The Spatial Depth Formulation

Unlike other contemporary work, this approach to medial axis estimation does not utilize the Delaunay or Voronoi decomposition. Rather, we use a localized spatial median [15] approach to estimating the medial axis. The advantage of the Voronoi decomposition is that it always provides the discrete medial axis, provided that the shape is perfectly known. The various techniques involving the Voronoi decomposition and sampled input data focus on different aspects of the problem of incompleteness. Our approach using the localized spatial depth formalization allows us to create an implicit interpolation of the medial axis. Areas without sufficient coverage that fail to create a ridge are thus passed over by the algorithm. Our reason for opting for the spatial median formulation is that it turns the discrete set of surface points into a suitable image which then allows us to take advantage of an array of computer vision techniques.

The initial input is the sampled surface points of an object. We begin by crafting a localized spatial depth image of the initial shape at a specified grid interval. The spatial depth of a surface scan is created by calculating a weight for each point p within an image based on a single observation (or surface scan point) o at a particular σ window size:

$$w(p,o) = e^{-\frac{(p-o)^2}{\sigma^2}}.$$

Once the weights have been computed at point p, we combine those weights of all surface scan points O into a spatial depth:

$$Spatial\,Depth(p,O) =$$
$$1 - \left\| \sum_j^n \left[\frac{w(p,O_j)}{\sum_{i=1}^n w(p,O_i)} \times \frac{p-O_j}{\sqrt{\sum_{i=1}^n (p-O_i)^2}} \right] \right\|.$$

The spatial depth is a reflection of the symmetry of a point in space to the surface points of our scan. The Spatial Depth formula has a bound of 0 to 1 and higher values indicate a better symmetry of surrounding surface points. It is by this property that we formulate our ridge detector.

2.1 Selecting the Right σ

Care should be taken in selecting the proper σ value in order to reconstruct the medial axis when using the spatial depth formulation in the context of estimating the the medial axis of a shape. Values of σ that are too small run the risk of being too local for a proper analysis of the ridges. Values of σ that are too large run the risk of creating a border area where the calculated values within σ units of the image edge are ineffective due to edge noise.

We created a cylinder shape to use as a test case for properly testing the most effective value of σ. The shape consists of a series of rings stacked to resemble that of a three dimensional cylinder. Each ring in the cylinder consist of 128 evenly spaced points along the perimeter of a circle with a radius of 1. The shape consists of 24 rings that are evenly spaced $\frac{1}{6}$ units apart. The overall shape consists of 3072 surface points. The overall dimension of the shape is 2 units by 2 units by 3.833 units.

A sufficiently small delta of 0.005 was selected and a spatial depth image was created using 100 σ values ranging from 0.01 to 2. The small delta produced an image dimension of 401 by 401 by 767 voxels. Only the center line of the cylinder image was relevant to our analysis. We plucked the single dimension line existing long the z-axis coordinate for all 100 σ values. Since the center line of the shape represents the highest conceptual spatial depth, we took the sum of each line. Our finding for this circumstance was that the σ representing the highest summation of spatial depths was 0.5, which is $\frac{1}{4}$ the diameter of our cylinder.

In practice, we feel that a σ should be selected that is roughly $\frac{1}{4}$ the diameter of tubular shape being studied. It may require observing the various appendages of the shape prior to analysis and determining a proper σ value.

3 Ridge Detection in 3D Space

In order to determine the ridges in three dimensional space, we must rotate each point within the (x, y, z) coordinate frame to a (p, q, r) coordinate frame individually so that one of the three directions is aligned with the ridge direction. To reduce the amount of noise in the image, a scale-space filter is first applied to the image. Scale space smoothing is a smoothing function using parameter t to determine the smoothness. t is a non-negative integer value which represents the scale of the smoothing. A larger t indicates a greater smoothing scale. If $t = 0$, the image is not smoothed. For an input samples with no noise, no smoothing is required. For noisy data, larger values of t will be required. The smoothing function is

$$g(x,y,z;t) = \frac{e^{\frac{-(x^2+y^2+z^2)}{2t}}}{2\pi t^{\frac{3}{2}}}.$$

The smoothed image is defined by the convolution of the smoothing function and the image function (a.k.a the brightness function):

$$L(x,y,z;t) = g(x,y,z;t) * f(x,y,z).$$

Lindeberg's 2D formulation of the scale space ridge detector required that each voxel within the image be rotated from a (x,y) coordinate frame to a (p,q) coordinate frame. The rotation matrix used to rotate the points was based on the eigenvectors discovered by the eigendecomposition of the Hessian matrix at point (x,y). In order to rotate each point within the image from the (x,y,z) coordinate frame to the (p,q,r) coordinate frame, we compute the 1st and 2nd derivatives of the image at point (x,y,z). The 1st derivatives are known as L_x, L_y, and L_z. From the 1st, we find the 2nd derivatives by taking the derivatives of each of the 1st derivatives. The 2nd derivatives are known as L_{xx}, L_{xy}, L_{xz}, L_{yy}, L_{yz}, and L_{zz}.

Using the 2nd derivatives of the image, we formulate the Hessian Matrix:

$$H_{(x,y,z)} = \begin{bmatrix} L_{xx} & L_{xy} & L_{xz} \\ L_{xy} & L_{yy} & L_{yz} \\ L_{xz} & L_{yz} & L_{zz} \end{bmatrix}.$$

Using the Hessian matrix, we perform the eigendecomposition to discover the eigenvalues of $H_{(x,y,z)}$: $\lambda_1, \lambda_2, \lambda_3$. The eigenvalues correspond to the 2nd derivatives of the system when rotated to the p, q, and r axis:

$$\begin{bmatrix} \lambda_1 \\ \lambda_2 \\ \lambda_3 \end{bmatrix} = \begin{bmatrix} L_{pp} \\ L_{qq} \\ L_{rr} \end{bmatrix}.$$

We also determine the eigenvectors for each of the eigenvalues,

$$V = \begin{bmatrix} x_1 & x_2 & x_3 \\ y_1 & y_2 & y_3 \\ z_1 & z_2 & z_3 \end{bmatrix},$$

where each column of V represents an eigenvector of the Hessian matrix.

We wish to formulate a rotation matrix that will transform the 1st derivatives of the image along the x, y, and z axis to the p, q, and r axis. We call this 3×3 rotation matrix R. If ∇g is the 1st derivatives at a point in (x,y,z) space, then $R \times \nabla g$ equals our 1st derivatives in (p,q,r) space, i.e.

$$\nabla g = \begin{bmatrix} \partial x \\ \partial y \\ \partial z \end{bmatrix},$$

$$R \times \nabla g = \begin{bmatrix} \partial p \\ \partial q \\ \partial r \end{bmatrix}.$$

We can show that

$$(R \times \nabla g) \times (R \times \nabla g)^T \times L$$

$$= \begin{bmatrix} L_{pp} & L_{pq} & L_{pr} \\ L_{pq} & L_{qq} & L_{qr} \\ L_{pr} & L_{qr} & L_{rr} \end{bmatrix}$$

$$= \begin{bmatrix} \partial p \partial p L & \partial p \partial q L & \partial p \partial r L \\ \partial p \partial q L & \partial q \partial q L & \partial q \partial r L \\ \partial p \partial r L & \partial q \partial r L & \partial r \partial r L \end{bmatrix}$$

$$= \begin{bmatrix} \lambda_1 & 0 & 0 \\ 0 & \lambda_2 & 0 \\ 0 & 0 & \lambda_3 \end{bmatrix}.$$

Now we evaluate $(R \times \nabla g) \times (R \times \nabla g)^T \times L$:

$$(R \times \nabla g) \times (R \times \nabla g)^T \times L$$
$$= (R \times \nabla g \times \nabla g^T \times R^T) \times L$$
$$= R \times (\nabla g \times \nabla g^T \times L) \times R^T$$
$$= R \times H_{(x,y,z)} \times R^T$$

By definition,

$$R \times H_{(x,y,z)} \times R^T = V^T \times H_{(x,y,z)} \times V.$$

Therefore, the matrix V^T is our rotation matrix R. With this we can determine L_p, L_q, L_r:

$$R \times \left[L_x L_y L_z \right]^T = \left[L_p L_q L_r \right]^T.$$

Using our values of L_p, L_q, L_r, L_{pp}, L_{qq}, and L_{rr}, we can define a ridge. A ridge in three dimensions is a point in space which has a negative gradient change along two 2nd derivatives and no negative gradient change along the third 2nd derivative. It's this third direction where a ridge line exists that can be followed for our purposes. We can define a ridge using the following:

$$
\begin{cases}
L_p = 0 \\
L_{qq} < 0 \\
L_{rr} < 0 \\
|L_{pp}| < |L_{qq}| \\
|L_{pp}| < |L_{rr}|
\end{cases}
\text{ or }
\begin{cases}
L_q = 0 \\
L_{pp} < 0 \\
L_{rr} < 0 \\
|L_{qq}| < |L_{pp}| \\
|L_{qq}| < |L_{rr}|
\end{cases}
\text{ or }
\begin{cases}
L_r = 0 \\
L_{pp} < 0 \\
L_{qq} < 0 \\
|L_{rr}| < |L_{pp}| \\
|L_{rr}| < |L_{qq}|
\end{cases}
$$

4   Implementing Canny's Edge Detector in 3D

To assist in determining which points can be identified as ridges, we implement a "dual threshold" technique for identifying ridges similar to the approach developed by Canny for identifying edges. Canny's edge detector required two things to be effective: strength and direction.

We formulate the strength of the ridge by summing the absolute values of the 2nd derivatives orthogonal to that of the ridge direction. Due to inevitable nuisances (i.e. rounding error), we must apply a narrow threshold on the ridge conditionals. We have found a threshold of 0.001 to be effective. We have developed a "ridge strength" (RS) to help us determine the strength of an edge based on the eigengap of a ridge in pqr-space:

$$
RS = \begin{cases}
|L_{rr}| + |L_{qq}| \\
\quad \text{if } |L_p| < ridgeThresh \wedge \\
\quad\quad L_{qq} < 0 \wedge L_{rr} < 0 \wedge \\
\quad\quad |L_{pp}| < |L_{qq}| \wedge |L_{pp}| < |L_{rr}| \\
|L_{pp}| + |L_{rr}| \\
\quad \text{if } |L_q| < ridgeThresh \wedge \\
\quad\quad L_{pp} < 0 \wedge L_{rr} < 0 \wedge \\
\quad\quad |L_{qq}| < |L_{pp}| \wedge |L_{qq}| < |L_{rr}| \\
|L_{pp}| + |L_{qq}| \\
\quad \text{if } |L_r| < ridgeThresh \wedge \\
\quad\quad L_{pp} < 0 \wedge L_{qq} < 0 \wedge \\
\quad\quad |L_{rr}| < |L_{qq}| \wedge |L_{rr}| < |L_{pp}| \\
0 \\
\quad \text{otherwise}
\end{cases}
$$

Provided that the RS score at a point is greater than 0, we determine the ridge direction RD by retrieving the eigenvector of the primary ridge determined by the maximum of L_{pp}, L_{qq}, or L_{rr}.

$$
RD = \begin{cases}
\begin{bmatrix} x_1 y_1 z_1 \end{bmatrix}^T, & \text{if } L_{pp} = max(L_{pp}, L_{qq}, L_{rr}), \\
\begin{bmatrix} x_2 y_2 z_2 \end{bmatrix}^T, & \text{if } L_{qq} = max(L_{pp}, L_{qq}, L_{rr}), \\
\begin{bmatrix} x_3 y_3 z_3 \end{bmatrix}^T, & \text{if } L_{rr} = max(L_{pp}, L_{qq}, L_{rr}).
\end{cases}
$$

It should be noted that while the input shape can and will contain holes, our algorithm makes no guarantees as to whether or not a discovered ridge exists on the interior or exterior of the shape. Still we can reduce the number of exterior ridges by filtering out all of the candidate ridges if they do not also have a local spatial depth greater than a high threshold (0.9 in our experiments).

Once we have determined the set of candidate ridges, their scores and their directions, we must now select two threshold values by which to select our ridges. We call these thresholds *high* and *low*. First, the image is scanned for all voxels with *RS* scores greater than *high*. These scores and locations are added to a priority queue where priority is determined by greatest *RS* score. While the priority queue is not empty, take the front element off the queue and label the point as a confirmed point. We then look at the two voxels in the directions of *RD* and $-RD$ and determine if those voxels have a *RS* score greater than *low*. If so, we add the voxel location and *RS* to the priority queue.

5 The Complete Algorithm

1 Create a localized spatial depth image I of the shape using a predetermined σ window size.
2 Perform the scale space smoothing filter on the image I with preferred scaling factor t.
3 For each point within the image I, determine the 1st which represents L_x, L_y, and L_z.
4 For each of the 1st derivatives, determine the 2st derivatives and formulate the Hessian matrix H.
5 Compute $[\lambda, V] = eig(H)$, where λ and V are the eigenvalues and eigenvectors of matrix H. The λ represents the 2nd derivative directions in the (p,q,r)-coordinate frame: L_{pp}, L_{qq}, L_{rr}.
6 Compute the rotation matrix R for this point, which is V^T.
7 Determine the 1st derivative directions in the (p,q,r)-coordinate frame, which is $R \times [L_x, L_y, L_z]^T$, which are represented by L_p, L_q, L_r.
8 For each point that meets the criteria for a ridge, compute the ridge score for this ridge point.
9 For each point that meets the criteria for a ridge, note the eigenvector associated with the largest eigenvalue of the Hessian matrix at this point. This becomes the dominate eigenvector.
10 Select a desired *low* and *high* threshold values for ridge points.
11 Create a three dimensional matrix equal to the size of the image I composed of all Boolean false flags. This matrix will represent the confirmed ridges.
12 For each ridge point that has a *RS* greater than the *high* threshold, add that ridge point's ridge score and location to the end of a priority queue.
13 While the priority queue is not empty:

  a Sort the priority queue by the ridge score in descending order.
  b Pop the first element from the priority queue. Flag this point in the confirmed ridge point matrix as true. If it has been previously flagged as a confirmed ridge, skip the remaining steps in this loop iteration.

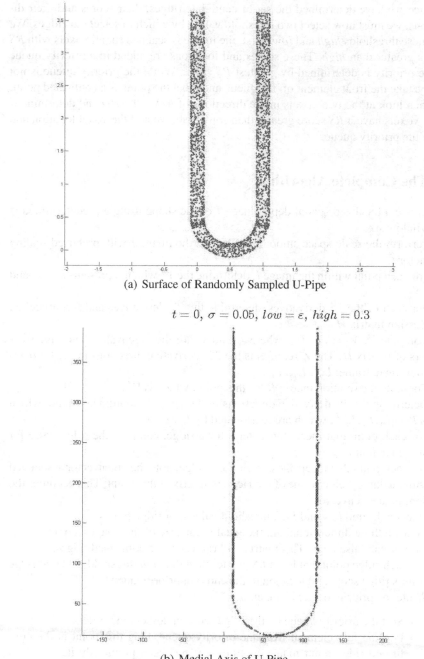

(a) Surface of Randomly Sampled U-Pipe

$t = 0,\ \sigma = 0.05,\ low = \varepsilon,\ high = 0.3$

(b) Medial Axis of U-Pipe

Fig. 1 Shape and Estimated Medial Axis

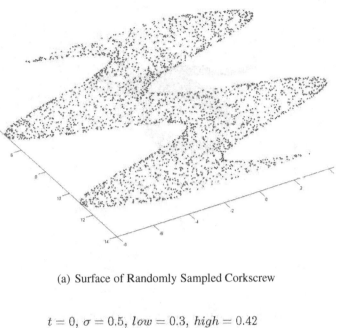

(a) Surface of Randomly Sampled Corkscrew

$t = 0, \sigma = 0.5, low = 0.3, high = 0.42$

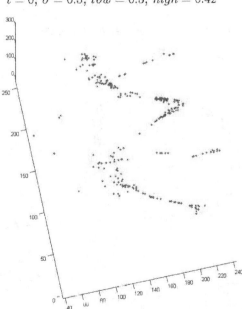

(b) Medial Axis of Corkscrew

Fig. 2 Shape and Estimated Medial Axis

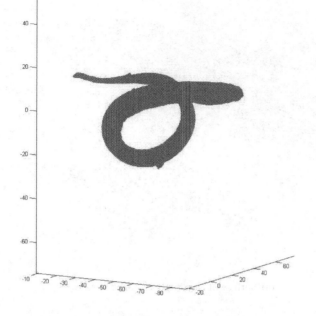

(a) Surface of Randomly Sampled Eel Surface

$t = 1,\ \sigma = 5,\ low = \varepsilon,\ high = 0.09$

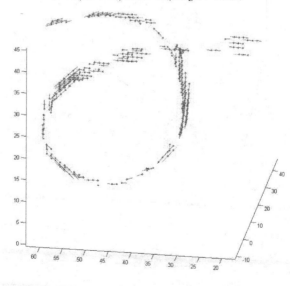

(b) Medial Axis of Eel

Fig. 3 Shape and Estimated Medial Axis

    c Test the neighboring point in the direction of the dominate vector for this point. If the neighboring point meets the criteria for a ridge and has a ridge score greater than the *low* threshold, add the neighboring ridge score and the ridge point location to the end of the priority queue.

    d Test the neighboring point in the direction opposite of the dominate vector for this point. If the neighboring point meets the criteria for a ridge and has a ridge score greater than the *low* threshold, add the neighboring ridge score and the ridge point location to the end of the priority queue.

6 Experimental Result

In order to test the validity of our algorithm, we created three dimensional images representing both artificial structures and real objects scanned with a 3D scanner. For each shape, we select an appropriate σ, smoothing parameter t, and ridge thresholds *high* and *low* and executed the algorithm. Examples of artificial images can been seen in Figures 1 and 2 and a surface image of an Anguilla rostrata (American Eel) in Figure 3. For the artificial structures, we evaluate each image based on the identified ridges by how closely they conform to a known center line of the shape. We have compared our results with the Power Crust method using a simple correspondence algorithm using a Root-Mean-Squared Error formulation comparing each point in the discovered medial axis to the nearest point in the ground truth (RMSE MA) and comparing each point the ground truth to the nearest point in the discovered medial axis (RMSE Truth). We can show that the two approaches are comparable in Table 1.

Table 1 We show that the Localized Spatial Depth and the Power Crust method have comparable results

U-Pipe	RMSE MA	RMSE Truth	Avg.
L. Spatial Depth	0.0786	0.2088	0.1437
Power Crust	0.0042	0.1117	0.0579
Cork Screw	RMSE MA	RMSE Truth	Avg.
L. Spatial Depth	1183.4	6748.3	3965.9
Power Crust	16182.0	2564.0	9373.1

7 Conclusions

In each experimental result, we have discovered enough ridge lines in order to connect the ridges using Dijkstra's shortest path algorithm. Our data sets mostly represent the elongated structures of fish with an anterior, a narrow body, and a posterior. By flagging the two points representing the anterior and posterior, we can use this technique to create a full length vertebra of the internal structure of the shape. After obtaining the body-length vertebra, to unfold the structure is trivial: associate each surface point with the nearest vertebra point and align it on a single axis, taking care to rotate the point with respect to the vertebrae point immediately preceding the connecting vertebra point. There is one problem with this approach which needs addressing in our future work. The body contortions of the fish specimen will cause the fish's body to twist around its own internal structure. The end result of our unfolding process will also be a twisted contortion around the discovered vertebra. We feel that the one way to alleviate this issue is to add landmarks along the dorsal side of the fish specimen. With this information, we can perform a rotation along the roll axis of the specimen.

Our algorithm depends heavily on the selection of a σ window. A good σ window can fall into a range of values, but it is still dependent on the diameter of the appendages of the object being analyzed. We feel that there is more research that can be done into evaluating the shape prior to the start of the algorithm that could be used in σ discovery.

Acknowledgment. This material is based upon work supported by the National Science Foundation under Grant No. NSF MCB-1027989 and MCB-1027830. We would like to thank Aishat Aloba for processing large data sets.

References

1. Chen, Y., Bart, H., Dang, X., Peng, H.: Depth-based novelty detection and its application to taxonomic research. In: Seventh IEEE International Conference on Data Mining, ICDM 2007, pp. 113–122. IEEE (2007)
2. Chen, Y., Bart Jr., H.L., Huang, S., Chen, H.: A computational framework for taxonomic research: diagnosing body shape within fish species complexes. In: Fifth IEEE International Conference on Data Mining, p. 4. IEEE (2005)
3. de Carvalho, M.R., Bockmann, F.A., Amorim, D.S., Brandão, C.R.F., de Vivo, M., de Figueiredo, J.L., Britski, H.A., de Pinna, M.C., Menezes, N.A., Marques, F.P., et al.: Taxonomic impediment or impediment to taxonomy? a commentary on systematics and the cybertaxonomic-automation paradigm. Evolutionary Biology 34(3), 140–143 (2007)
4. Rodman, J.E., Cody, J.H.: The taxonomic impediment overcome: Nsf's partnerships for enhancing expertise in taxonomy (peet) as a model. Systematic Biology 52(3), 428–435 (2003)
5. Bajaj, C.L., Bernardini, F., Xu, G.: Automatic reconstruction of surfaces and scalar fields from 3d scans. In: Proceedings of the 22nd Annual Conference on Computer Graphics and Interactive Techniques, pp. 109–118. ACM (1995)
6. Blum, H.: A transformation for extracting new descriptors of shape. Models for the Perception of Speech and Visual Form 19(5), 362–380 (1967)
7. Boissonnat, J.-D.: Geometric structures for three-dimensional shape representation. ACM Transactions on Graphics (TOG) 3(4), 266–286 (1984)
8. Edelsbrunner, H., Mücke, E.P.: Three-dimensional alpha shapes. ACM Transactions on Graphics (TOG) 13(1), 43–72 (1994)
9. Amenta, N., Choi, S., Kolluri, R.K.: The power crust. In: Proceedings of the Sixth ACM Symposium on Solid Modeling and Applications, pp. 249–266. ACM (2001)

10. Dey, T.K., Goswami, S.: Tight cocone: a water-tight surface reconstructor. In: Proceedings of the eighth ACM Symposium on Solid Modeling and Applications, pp. 127–134. ACM (2003)
11. Chazal, F., Lieutier, A.: The λ-medial axis. Graphical Models 67(4), 304–331 (2005)
12. Lindeberg, T.: Edge detection and ridge detection with automatic scale selection. International Journal of Computer Vision 30(2), 117–156 (1998)
13. Sato, Y., Nakajima, S., Shiraga, N., Atsumi, H., Yoshida, S., Koller, T., Gerig, G., Kikinis, R.: Three-dimensional multi-scale line filter for segmentation and visualization of curvilinear structures in medical images. Medical Image Analysis 2(2), 143–168 (1998)
14. Canny, J.: A computational approach to edge detection. IEEE Transactions on Pattern Analysis and Machine Intelligence 6, 679–698 (1986)
15. Chen, Y., Dang, X., Peng, H., Bart, H.L.: Outlier detection with the kernelized spatial depth function. IEEE Transactions on Pattern Analysis and Machine Intelligence 31(2), 288–305 (2009)

Pipelined Multi-GPU MapReduce for Big-Data Processing

Yi Chen, Zhi Qiao, Spencer Davis, Hai Jiang, and Kuan-Ching Li

Abstract. MapReduce is a popular large-scale data-parallel processing model. Its success has stimulated several studies of implementing MapReduce on Graphic Processing Unit (GPU). However, these studies focus most of their efforts on single-GPU algorithms and cannot handle large data sets which exceed GPU memory capacity. This paper describes an upgrade version of MGMR, a pipelined multi-GPU MapReduce system (PMGMR), which addresses the challenge of big data. PMGMR employs the power of multiple GPUs, improves GPU utilization using new GPU features such as streams and Hyper-Q, and handles large data sets which exceeds GPU and even CPU memory. Compared to MGMR, the newly proposed scheme achieves a 2.5-fold performance improvement and increases system scalability, while allowing users to write straightforward MapReduce code.

Keywords: MapReduce, big-data, multi-GPU, stream, concurrency, Hyper-Q.

1 Introduction

Large-scale data processing is tougher than ever before since the size of the data is increasing too fast for hardware resources to keep up. Nowadays, Graphics Processing Units (GPU) are widely used in the High Performance Computing world to enhance job throughput, as its architecture is quite data-parallel friendly.

Yi Chen · Zhi Qiao · Spencer Davis
Dept. of Computer Science, Arkansas State University, USA
e-mail: {yi.chen,zhi.qiao,spencer.davis}@smail.astate.edu

Hai Jiang
Dept. of Computer Science, Arkansas State University, USA
e-mail: hjiang@astate.edu

Kuan-Ching Li
Dept. of Computer Science & Information Eng., Providence University, Taiwan
e-mail: kuancli@pu.edu.tw

R. Lee (Ed.): *Computer and Information Science*, SCI 493, pp. 231–246.
DOI: 10.1007/978-3-319-00804-2_17   © Springer International Publishing Switzerland 2013

MapReduce has also been widely adopted to solve Big Data problems[6]. It was originally proposed by Google to pursue a simple and flexible parallel programming paradigm. With MapReduce, users need only write *Map* and *Reduce* functions to solve problems in parallel. The underlying programming details, such as how to handle communication among data nodes, are transparent to users. Data affinity across the network and fault tolerance among multiple nodes can be achieved automatically.

Our previous work, MGMR, has already migrated the MapReduce framework into the GPU environment and successfully utilized Multiple GPUs concurrently [4]. As GPU technology advances, more powerful GPUs such as NVIDA Kepler have been developed. To leverage the most advanced features of Kepler such as Asynchronous Dual-channel Data Transfer and Hyper Q, we introduce a pipelined workflow into MGMR to make our system more powerful and efficient.

In order to handle Big Data sets, increasing only the computability is not enough. To push our system even further, we use the idea of Memory Hierarchy in MGMR and enhance its ability to store the huge amount of data as well as efficiently transfer data between disks and processing units. This paper proposes a Pipelined Multi-GPU MapReduce implementation for Big-Data Processing, called PMGMR, and makes the following contributions:

- Multiple GPUs are utilized to accelerate MapReduce operations.
- Pipelined workflow maximizes the usage of the GPU by overlapping communication and computation. Since we are hiding the traffic behind computation, PMGMR minimizes the overall communication time, thereby increasing the throughput.
- A job scheduler is employed to manage memory and data-flow. Input size is no longer bounded by GPU memory and even CPU memory.
- For Fermi architecture, a GPU operation scheduler is implemented to stabilize performance when GPU operations are issued from multiple CPU threads.
- A configuration optimizer is adopted, which will collect the running result to dynamically adjust the environment settings and ensure a balanced load. As the computation continues, the result will move closer to the ideal curve.

The remainder of this paper is organized as follows: Section 2 introduces necessary GPU architecture and MapReduce framework background. Section 3 will explain the design of PMGMR system and how the efficiency is achieved. We will discusses our experimental results in Section 4 by comparing our Pipelined system with other similar systems. In Section 5, some related MapReduce implementations are briefly discussed. Finally, the conclusion and future work are given in Section 6.

2 Background

PMGMR was developed in CUDA targeting Nvidia Fermi and Kepler architectures.

2.1 Big Data

Big Data is a collection of large and complex data sets which are difficult to be processed by relational database management systems or traditional data processing applications. The challenges with Big Data include capture, storage, search, analysis, etc. The typical size of a Big Data problem is in terabytes scale as of 2012, and the size is constantly growing.

The explosion of Cloud Computing and mobile networks boost the need for Big Data processing. Large companies like Amazon and Google tend to keep track of and analyze users' information and browsing habits to improve the customer experience[1]. Companies like Twitter need to process hundreds of millions of tweets per day and search keywords to identify emerging trends. All of these companies are facing the situation that data size grows exponentially. Therefore, their top priority job is not just to have huge storage space but also to process this information as quickly as possible as real time response becomes vital.

2.2 Multi-GPU Architecture

Each NVIDIA GPU consists of multiple streaming-multiprocessors (SMs) that can execute thousands of light-weight hardware threads concurrently. CUDA helps map thread hierarchy onto GPU cores. Up to 512 threads are grouped into thread blocks which are then assigned to SMs. Each SM then schedules work in groups of 32 parallel threads, called warps. Extremely fast context switch between warps can help tolerate memory access latency. All threads within the same block can access the common shared memory. This helps synchronize threads within the same block, and facilitate extensive reuse of on-chip data in order to greatly reduce off-chip traffic.

For a machine with multiple NVIDIA Fermi GPUs, GPUDirect is a technique used to handle inter-GPU communication via the PCIe bus directly without CPU side data buffering[13]. High-speed DMA (Direct Memory Access) engines enable this inter-GPU communication. With GPUDirect, one GPU can directly read/write data from/to another GPU's memory within the same machine. By eliminating unnecessary copies in system memory (on CPU side), Fermi GPU achieves significant performance improvement in data transfer. Asynchronous bidirectional memory copy is another advanced feature that not only doubles the data transfer bandwidth, but also helps achieve the overlapping of computation and communication.

For the latest Kepler GPU, GPUDirect can even transfers between 3rd party devices such as a Network Card. This feature made direct communication between Kepler GPUs across many machines possible.

2.3 MapReduce Programming Model

MapReduce is a programming model for processing large data sets. MapReduce is widely used in various domains such as machine learning, data mining[11], and bioinformatics[7]. Moreover, the MapReduce model has been adapted to many computing environments like multi-core system[8, 14], desktop grids[9, 5], and mobile platforms. The design goals of MapReduce include programmability, robustness and scalability.

Writing a parallel-executable program has been proven over the years to be very challenging. MapReduce provides regular programmers a much easier way to produce paralleled distributed programs, by requiring them to write only simpler *Map* and *Reduce* functions, which focus on the logic of the specific problem, while the MapReduce system automatically takes care of the underlying details such as parallelism, communication, fault tolerance, and load balancing.

The *Map* and *Reduce* functions are both defined with respect to data structured in (key, value) pairs. *Map* takes one pair of data in one data domain, and returns a list of pairs in a different domain:

$$\mathbf{Map} : (k_1, v_1) \rightarrow list\,(k_2, v_2) \tag{1}$$

The *Map* function is applied in parallel to every pair in the input dataset. This produces a list of pairs for each call. After *Map* stage, the MapReduce system collects all pairs with the same key from all lists and groups them together, creating one group for each key. This stage typically called *Shuffle* or *Partition*.

The *Reduce* function is then applied in parallel to each group. This in turn produces a collection of values in the same domain:

$$\mathbf{Reduce} : (k_2, list\,(v_2)) \rightarrow list\,(k_3, v_3) \tag{2}$$

Each *Reduce* call typically produces either one value or an empty return, though one call is allowed to return more than one value. The returns of all calls are collected as the desired result list.

Thus MapReduce system transforms a list of (key, value) pairs into a list of values. This behavior is different from the typical functional programming map and reduce combination, which accepts a list of arbitrary values and returns one single value that combines all the values returned by map.

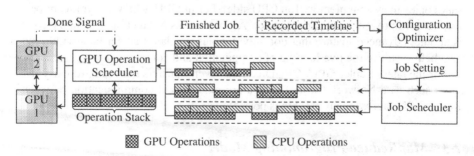

Fig. 1 The overview of PMGMR work-flow in Fermi architecture

3 PMGMR System Design

PMGMR mainly consists of three functional units: a job scheduler, a GPU operation scheduler, and a configuration optimizer. They are design to improve MGMR [4] in the following respects: capability of handling large data sets, stability of GPU efficiency, and performance of the overall system. Fig. 1 shows the overview of PMGMR work-flow. The job scheduler launches Map and Reduce jobs based on setting. For Fermi architecture, GPU operations inside these jobs are emitted to GPU operation scheduler and issued in optimized order. After each job finishes its execution, the recorded timeline is sent to the configuration optimizer, and job setting is adjusted at runtime according to the analysis result.

3.1 Challenges of Big Data and Our Approach

Big data has already become a challenge for MapReduce, just as in other data processing systems[12]. A common GPU-based MapReduce application tends to read input from the hard disk, and then MapReduce system will transfer the data to CPU memory and finally GPU memory. Under the current storage hierarchy, the capacity of the hard drive is much bigger than CPU memory, and this gap also exists between CPU memory and GPU memory[10, 2]. Thus, large input can easily cause overflow with GPU and even CPU memory.

PMGMR is designed to process massive amounts of data on each computing node. When the processed data does not fit GPU and CPU memory, PMGMR employs a job scheduler to handle the data transfer among hard drive, CPU memory, and multiple GPUs. The job scheduler ensures that no memory overflow will occur due to large input. This frees the user to import any size of data without defining extra functionalities, as long as the data and its output can reside in the hard disks.

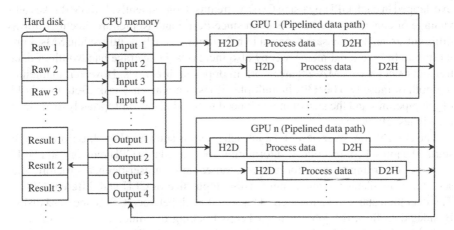

Fig. 2 The data-flow overview in one cycle

The data scheduler of PMGMR typically uses a divide-and-conquer strategy which separates input data into smaller pieces and merges their outputs back later. This strategy applies to both Map and Reduce stages because of the following reason: The input of Map stage can be partitioned into chunks of any size, and the input of Reduce stage is only indivisible for values with the same key. The implementation of Shuffle stage is inspired by the idea of external sorting, but instead of using CPU, the data is transferred and sorted in multiple GPUs.

The data-flow overview of data scheduling is show in Fig. 2. Part of the data is loaded from the hard disks into CPU memory. Then this partition is further logically divided into smaller chunks which fit GPU memory. The scheduler keeps GPUs busy with input chunks and transfers their outputs back to CPU memory. After all chunks are processed, the outputs of the current partition are combined and written into hard disks. This process is repeated until all parts of the input in the hard drive are processed.

3.2 Pipelined Multi-GPU Utilization

Full GPU utilization is hard to achieve. For an ordinary user, designing efficient Map and Reduce functions in a GPU-based MapReduce system is always difficult. Even with the well-designed official library from Nvidia, the average utilization of a single kernel is still around 80 percent which means roughly 20 percent of the computation power is wasted.

PMGMR employs multiple GPUs to process big data with tremendous throughput. Meanwhile, instead of relying on the user, the runtime scheduler dramatically advances the efficiency of each Nvidia GPU by using CUDA streams and Hyper-Q. CPU idle time is also slashed by allowing multiple CPU cores to simultaneously utilize multiple GPUs.

The overview of the pipelined data-path is shown in Fig. 2. The data-paths are overlapped in each GPU by using CUDA streams. Consequently, PMGMR takes advantage of concurrent kernel execution since Fermi and Kepler architectures allow different streams of the same context to run simultaneously to utilize idle SMs. Furthermore, bidirectional memory copy also increases the average bandwidth of data transfers since two DMA engines work in opposite directions. However, maintaining both of these two benefits in multiple GPUs is not an easy task. Both threshold of job pipelines and the size of input all need to be adjusted appropriately to achieve the highest efficiency.

The control-flow overview of the job scheduler is shown in Fig. 3. Before the job starts, the job scheduler collects information about available CPU memory and GPU devices. According to the information gathered and the user-defined structure, the scheduler calculates the maximum feasible input size of CPU and different GPUs. The data then starts to be transferred back and forth between GPUs and hard disks. Besides data-flow, the job scheduler keeps tracking the timeline of each job and optimizes the job setting based on the previous records. The goal of this runtime

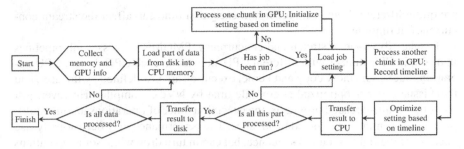

Fig. 3 Control flow of the job scheduler

optimization is to continuously improve the overall efficiency of multiple GPUs by approaching the optimal threshold and input size.

3.3 Fermi Pipelining Schemes

Streams and concurrency were implemented in CUDA 4.0 with the Fermi architecture. With GPUs which have compute capability not less than 2.0, a programmer is able to launch up to 16 concurrent CUDA kernels, and two bidirectional *cudaMemcpyAsync()* calls. As defined in the Fermi architecture white paper, the Fermi hardware has one compute engine queue and two copy engine queues. Stream dependencies between engine queues are maintained, but become FIFO sequence within an engine queue. A CUDA operation is dispatched from the engine queue if preceding calls in the same stream have completed, or if preceding calls in the same queue have been dispatched.

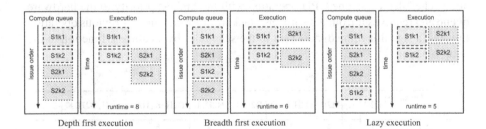

Fig. 4 Different runtime with different execution orders

By appropriately using streams to achieve operation overlap and high GPU occupancy, some applications which previously used default stream, thereby limiting GPU utilization, show several times performance improvement with the same algorithm and only minor code changes. On the other hand, a blocked operations stops all other operations in the queue, even in other streams. For instance, as shown in Fig. 4, two streams issue 4 kernels which have different execution times. Each kernel fills 50 percent of the SM resources. The result shows that their overall runtimes are quite different. Both issue order and execution time can affect the stream concurrency at runtime.

To take advantage of stream and concurrency, PMGMR processes job pipelines in multiple CPU threads with different GPU streams. In each CPU thread, GPU operations such as memory copy and kernel execution are pre-defined by the user, and their issue order is optimized at compile time by NVCC compiler. However, just like other multi-threaded programs, the execution order of operations among multiple threads cannot be predicted in advance. The performance may occasionally rise because of the right execution sequence, but can in turn drop while some operations block others. This problem is mainly caused by the limitations of the Fermi architecture: Only one FIFO hardware queue exists for all the software streams. Thus the GPU operations need to be issued wisely to avoid unnecessary blocking in the hardware queue.

This situation is handled by the runtime scheduler for the Fermi architecture. Instead of launching a kernel function directly inside the mapper or reducer, the user calls our wrap function to pass their execution plans to the runtime scheduler. Then, the issue order of these plans are reordered and dispatched by the scheduler with a reorder-and-fire scheme.

3.3.1 Reorder-and-Fire Scheme

The goal of our reorder-and-fire scheme is to significantly reduce the operation combination which causes blocking. Instead of directly issuing GPU operations, the execution plan of each operation is first sent to the GPU operation scheduler and placed into a software stack with a limited window size. The priorities of these operations are then assessed by the scheduler depending on the operations that remain in the GPU. When the pending operation in the stack exceeds its window size, GPU operation scheduler sends one GPU operation with the highest priority to the hardware queue and removes it from the software stack. Moreover, when a previous operation finishes, a signal is immediately sent to the scheduler by a callback function in the wrap function.

Coming \ Remained	D2H	K	H2D	D2H, H2D	D2H, K	H2D, K	D2H, K, H2D	D2H, D2H, K	......
D2H	1	2	2	3	3	4	5	4	......
K	2	0|2	2	4	2	2	6	4	......
H2D	2	2	1	3	4	3	5	6	......

D2H: Device-to-Host   K. Kernel   H2D: Host-to-Device

Fig. 5 Priority table of coming GPU operations when proportions of Device-to-Host, Kernel, and Host-to-Device are the same

Fig. 5 shows part of the priority table for different combinations of coming operation and GPU scenes when proportions of Device-to-Host, Kernel, and Host-to-Device operations are the same. These priorities need to be adjusted for different operation proportions which represent the chances of their executions. Fig. 5 does not include the situations where GPU operations are from the same stream. PMGMR will not issue the next operation of the same stream until the current one is finished. This strategy is part of our reorder-and-fire scheme because operations of the same stream are always serialized, and issuing them in advance does not increase concurrency but only raises the probability of blocking others.

Because of the restriction in the Fermi architecture, the possible operation combinations which create concurrency are also limited. Based on feasible combinations, an efficient scheme is designed to evaluate the priorities of different combinations which describe the chance to increase concurrency and avoid blocking. As shown in Fig. 5, the foundation priorities are the nine pairs which have one coming and one remaining operation. These nine priorities are defined according to these four situations:

1. *cudaMemcpyAsyncs()* in the same direction are serialized (1 credits)
2. Bidirectional *cudaMemcpyAsyncs()* are overlapped (2 credits)
3. Sequentially issued kernels delay signals and block *cudaMemcpyAsyncs()* (0 credits)
4. Kernels in different streams can be executed concurrently when resources are available. (2 credits)

In situation 1, although memory copies are serialized, kernels and memory copies in the opposite direction can still run concurrently aside. Both situation 2 and 4 are the perfect cases to increase concurrency. Situation 3 is a very special case and actually the worst case which shows that inappropriate concurrent kernel execution can also block other operations. The solution is to insert memory copy between the sequential kernels. These situations represent four unique circumstances, and only situation 3 relies on issue order. Thus, for the GPU scene which has more than one remaining operation, its priorities can be evaluated by adding all priorities of its one-to-one pairs together while distinguishing two different cases of situation 3. The generalized scheme is described in Fig. 6.

3.3.2 Execution Plan Queue

As a prerequisite of the previous schemes, the runtime scheduler needs to have the capability of reordering and issuing the user's execution plans. This goal has two main challenges:

1. The definition of the user's kernel function is unknown.
2. User-defined GPU operations from different streams need to be reordered at runtime.

To address these challenges, PMGMR employs the most recent version of the standard C++ programming language, C++11, which includes several additions to the core language and extends the C++ standard library. With a new feature of C++11, user-defined functions can be stored as an object which is movable and copyable. Therefore, the wrap function can simply be sent together with its operation type to

---

Algorithm 1 Operation priority formula

1: **procedure** PRIORITY(*coming, remain*)   ▷ Coming and remaining operations
2:     *total* ← 0
3:     *count* ← *amount*(*remain*)
4:     **for** $i = 1 \rightarrow count$ **do**
5:         *this* ← *checkPair*(*coming, remain*[*i*])   ▷ Priority of each operation pair
6:         **if** $i = count$ **then**
7:             **if** *coming* = *K* & *remain*[*i*] = *K* **then**   ▷ K represents Kernel
8:                 *this* ← 0
9:             **end if**
10:        **end if**
11:        *total* ← *total* + *this*
12:    **end for**
13:    **return** *total*   ▷ Total priority of all pairs
14: **end procedure**

Fig. 6 Pseudocode for operation priority calculation

the runtime scheduler. According to its type, the scheduler can now evaluate their priorities and issue them in an optimized order.

However, most of features in C++11 requires at least gcc 4.7 or g++ 4.6 as a compiler. Since the NVCC compiler (CUDA's compiler) uses gcc 4.6 as its default host compiler, the normal compilation fails because of the unsupported syntax. Nevertheless, by separating the CUDA and C++11 codes into different files, the codes are first compiled separately with NVCC and g++ and linked back to an executable file later.

3.4 Kepler Pipelining Scheme

Kepler improves concurrency functionality with the new Hyper-Q feature. Hyper-Q increases the total number of connections (work queues) between the host and the CUDA Work Distributor logic in the GPU by allowing 32 simultaneous, hardware-managed connections (compared to the single connection available with Fermi). Therefore, up to 32 streams can run totally dependently in Kepler GPU.

PMGMR takes great advantage of Hyper-Q because job pipelines are processed in different CPU threads. Since GPU operations in different GPU streams in Kepler are actually maintained in different hardware connections, PMGMR no long need to take care of the cross dependency issue which previously encountered false serialization across tasks in Fermi. Thence, the GPU utilization is always high and stable because concurrency and bidirectional memory transfer is easy to achieve in Kepler.

3.5 Runtime Tuning and Load Balancing

For the job scheduler, job pipelines are issued according to two options in job setting: threshold of job pipelines and the input size of each job. These two options are adjusted by the configuration optimizer with timeline records and memory overflow detection.

The threshold option represents the number of running CPU threads which process input data with user-defined *Map* and *Reduce* functions. If the number is too small, concurrency cannot be high enough to fully utilize multiple GPUs. However, if too many job pipelines are generated, the performance can also decrease because of thread context switch latency. PMGMRs adjust this option based on the processing throughput. Initially, the number of threshold is small. Assume the initial number is N. After N job pipelines are launched, the configuration optimizer increases this number by one, and keeps tracking the trend of the processing throughput. After these N+1 job pipelines are finished, if the average job throughput still increases, the number will again be increased by one. The process is repeated until the average throughput starts to drop.

The input size option also needs to be adjusted carefully to maximize the usage of GPU memory, while still avoids blocking operations because of memory overflow detection. Initially, the size of input and its output is equally to *GPUMemory/PipelineNumber*, and no memory overflow can occur in this situation. For the purpose of maximizing GPU occupancy, the configuration optimizer gradually increases the input size after the threshold becomes stable, while the possibility of memory overflow also increases. Thus before any GPU operation is issued, the wrap function checks the GPU memory usage and decide whether this operation can be issued or not. If one operation may overflow GPU memory, its job pipeline will be blocked until it is safe to issue this operation. The performance could drop because of the blocked CPU operations and lower GPU utilization. Therefore, when job scheduler detects a memory overflow operation, the option input size is rolled back to the previous state, and the maximum feasible GPU utilization is reached for this job.

4 Experimental Results

All experiments were conducted on two servers. One machines contains two Intel Xeon E5504 (2.00GHz, totally 8 cores) with 24 GB RAM and two Nvidia Tesla C2050 (1.15 GHz, 448 CUDA Cores, 2,687 MB global memory). Another machine contains four Intel Xeon E5-2620 (2.00GHz, totally 24 cores) with 32 GB RAM and two Nvidia Tesla K20Xm (0.73 GHz, 2688 CUDA Cores, 5,760 MB global memory). Both servers are running the GNU/Linux operating system with kernel version 2.6.32. Testing applications are implemented with C++11 and CUDA 5.0 and compiled with g++ and NVCC compiler in CUDA Toolkit 5.0.

4.1 Pipelining Scheme

The pipelining scheme of PMGMR utilizes multiple GPUs mainly in two ways: concurrent kernel execution and bidirectional memory copy. K-Means Clustering (KMC) is used in data mining which aims to partition n observations into k clusters where all observations in a cluster are close to the nearest mean. We use KMC to test the pipelining scheme because of its computation-heavy characteristic. Since concurrent kernel execution is originally designed for inefficient kernels, if the performance of KMC can still be improved by pipelining scheme, then this scheme can perform even better in other benchmarks with inefficient kernels which originally waste the computation power of multiple GPUs.

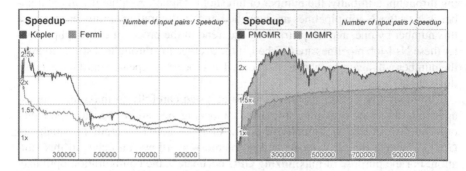

Fig. 7 Two speedups: The left one is PMGMR's runtime speedup over MGMR; The right one is the speedup of Kepler version of both PMGMR and MGMR over Fermi version

As shown in Fig. 7, the pipetting scheme has considerable speed-up over the sequential execution when the number of input pairs are small. As the number grows bigger, the speedup become smaller in both Kepler and Fermi versions because each kernel has already consumed most of GPU resources. However, the execution time is still reduced by asynchronous memory copy because most of the memory copying are overlapped with computations.

Kepler gains more benefits from the pipelining scheme than Fermi does. The reason is that both hardware connections and level of concurrency are improved in Kepler. As a result, when the input and output data sets can fit GPU memory, the Kepler version of PMGMR achieves 1.8 times speed-up over Fermi version, while the Kepler version of MGMR only achieves 1.6 times speed-up.

4.2 Overall Performance

A 60GB binary file which contains the original coordinate information of KMC, is generated for measuring the overall performance of PMGMR. Since KMC is NP-hard, we set the maximum rounds to 3 for measuring performance. In this test, the first several rounds has relatively low execution time. After certain size of input, and actually before this size can overflow CPU memory, the execution time starts to increase dramatically in the coming rounds. Then, after few rounds, the execution time becomes stable again with a much lower throughput. The main reason of this is that Linux maintains caches for part of the file system. Originally, most of data that PMGMR reads are from the caches which resides in CPU memory. After all the cache are flushed out from CPU memory because of the memory operations, PMGMR starts to read data from hard drive which has very slow read/write speed and actually becomes a main bottleneck.

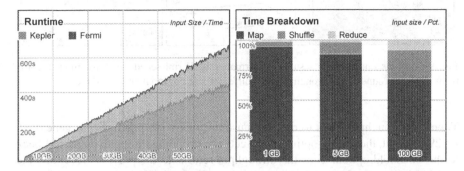

Fig. 8 K-means Clustering experimental results: execution time and runtime breakdown

As shown in Fig. 8, notice that when the input size is roughly less than 3GB, the execution time can almost be ignored if we compare it with the following execution time. After that, the performance of PMGMR is mainly bounded by hard disk speed. Since the read/write combined speed of hard drive is 105MB/s in Fermi machine and 163MB/s in Kepler machine, PMGMR shows very low operation overheads while continuously processing input and storing output. The difference of execution time between Fermi and Kepler machines is also caused by different hard drive speeds. Moreover, the runtime breakdown shows that as the input size increases, the proportion of shuffle and reduce stages do not increases. The reason is that a *Partial Reduce* stage is used for each portion to reduce I/O. Thus, only the sum of the x-y coordinates of each cluster in each portion is written into the output file of the *Map* stages. Since the number of clusters is limited, total size of the output files is always smaller compared to the input of the Map stage. Thus *Shuffle* and *Reduce* stages become very light-weighted.

4.3   Runtime Optimization

A GPU operation scheduler and a setting optimizer are employed to improve the stability and throughput in PMGMR. The first chart in Fig. 9 shows the execution time comparison between normal multi-threaded Fermi version and scheduled multi-threaded Fermi version. The experimental result shows that compared to the normal Fermi version, the scheduled one is much stabler. Although sometimes the normal Fermi scene shows shorter execution time, the average performance of the scheduled version is still higher.

The setting optimizer mainly uses two steps to optimize settings. Experimental results show that setting optimizer works well as it is designed. As shown in the third chart in Fig. 9, the throughput roughly keeps increasing until the 22nd round.

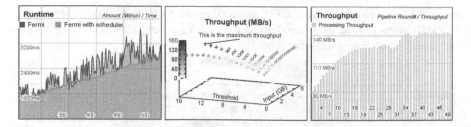

Fig. 9 Runtime optimization results: performance of Fermi scheduler, throughput records from different rounds, and throughput with different sizes of input and threshold

Then the throughput starts to drop because too many job pipelines are generated. At 29th round, the setting optimizer detects the dropping throughput and roll back to the threshold option. Thus the throughput returns to the previous level and the threshold option is fixed. After that, the size of input is gradually increased every round until 36th round. A memory overflow operation is detected and blocked by the job scheduler. Therefore, the option of input is also rolled back to the previous state. Finally, the throughput becomes stable, and both options are finalized for the current MapReduce stage.

5 Related Work

MapReduce has been implemented on many different platforms such as computer cluster, shared memory system, and CPU-GPU coupled architecture. We implement PMGMR based on the contributions and some problems which are found in the following MapReduce implementations.

Hadoop MapReduce [15] developed by Apache Software is designed for better programmability in processing vast amount of data in clusters. Hadoop was developed in Java, but Hadoop Streaming allows users to customize their own Map and Reduce functions in other programming languages such as C and python. Hadoop Streaming is one of future branches for testing PMGMR in cluster environment besides MPI.

GPMR [14] extends GPU MapReduce to GPU cluster level. By splitting the large input of map and reduce task into chunks and using partial reductions and accumulation which reduce network IOs, GPMR pushes GPU clusters to achieve the speedup over other MapReduce libraries. However, GPMR only focus on single GPU per node, neglecting that communication between multiple GPUs on the same node can be much cheaper. Moreover, the performance of GPMR may vary by the characteristics of different tasks. Hence each job needs extra configuration to achieve its best potential.

A Coupled CPU-GPU MapReduce [3] implements two schemes which use both CPU and GPU to process data. In Map-Dividing scheme, the data of Map and Reduce stage is processed in both CPU and GPU. In Pipelining Scheme, each device is only responsible for doing one stage of the MapReduce. To decrease load imbalance of GPU and CPU, the runtime system reduces the block size of each worker in order to reduce the difference among the finish times of the workers. However, this technique could also decrease GPU utilization when the size of input is to small to leash GPU power.

6 Conclusions and Future Work

Previous efforts have noted the importance of adapting to big data issues while designing a data processing tool. PMGMR allows users to import data which exceeds GPU and CPU memory. A pipelining scheme and a runtime optimizer are designed to maximize GPU utilization and increase processing throughput. In Fermi architecture, the problem of unstable performance with multiple CPU threads and streams, is addressed by a GPU operation scheduler. Experimental results have shown PMGMR's advantage over MGMR in all capability, performance, and stability.

The future work includes extending PMGMR to GPU Clusters by using RDMA and Hyper-Q for further performance scalability, integrating it with distributed file systems for fault tolerance, and improving its easy-to-use aspect with the newest c++11 standard for programmability.

References

1. Bollier, D., Firestone, C.M.: The promise and peril of big data. Aspen Institute, Communications and Society Program (2010)
2. Chen, L., Agrawal, G.: Optimizing mapreduce for gpus with effective shared memory usage. In: Proceedings of the 21st International Symposium on High-Performance Parallel and Distributed Computing, pp. 199–210 (2012)
3. Chen, L., Huo, X., Agrawal, G.: Accelerating mapreduce on a coupled cpu-gpu architecture. In: Proceedings of the International Conference on High Performance Computing, Networking, Storage and Analysis, p. 25 (2012)
4. Chen, Y., Qiao, Z., Jiang, H., Li, K.C., Ro, W.W.: Mgmr: Multi-gpu based mapreduce. In: To Appear in Proceedings of the 8th International Conference on Grid and Pervasive Computing (2013)
5. Czajkowski, K., Fitzgerald, S., Foster, I., Kesselman, C.: Grid information services for distributed resource sharing. In: Proceedings of 10th IEEE International Symposium on High Performance Distributed Computing, pp. 181–194 (2001)
6. Dean, J., Ghemawa, S.: Mapreduce: Simplied data processing on large clusters. Communications of the ACM 51(1), 107–113 (2008)
7. Dinov, I.D.: Cuda optimization strategies for compute-and memory-bound neuroimaging algorithms. Computer Methods and Programs in Biomedicine (2011)
8. Fadika, Z., Dede, E., Hartog, J., Govindaraju, M.: Marla: Mapreduce for heterogeneous clusters. In: Proceedings of the 2012 12th IEEE/ACM International Symposium on Cluster, Cloud and Grid Computing, pp. 49–56 (2012)
9. Foster, I., Kesselman, C.: The grid 2: Blueprint for a new computing infrastructure. Morgan Kaufmann (2003)

10. Ji, F., Ma, X.: Using shared memory to accelerate mapreduce on graphics processing units. In: Proceedings of the IEEE International Parallel & Distributed Processing Symposium, pp. 805–816 (2011)
11. Jinno, R., Seki, K., Uehara, K.: Parallel distributed trajectory pattern mining using mapreduce. In: IEEE 4th International Conference on Cloud Computing Technology and Science, pp. 269–273 (2012)
12. Nakada, H., Ogawa, H., Kudoh, T.: Stream processing with bigdata: Sss-mapreduce. In: Proceedings of 2012 IEEE 4th International Conference on Cloud Computing Technology and Science, pp. 618–621 (2012)
13. Shainer, G., Lui, P., Liu, T.: The development of mellanox/nvidia gpu direct over infiniband a new model for gpu to gpu communications. In: Proceedings of the 2011 TeraGrid Conference: Extreme Digital Discovery, vol. 26, pp. 267–273 (2011)
14. Stuart, J.A., Owens, J.D.: Multi-gpu mapreduce on gpu clusters. In: Proceedings of the 2011 IEEE International Parallel & Distributed Processing Symposium, pp. 1068–1079 (2011)
15. White, T.: Hadoop: The Definitive Guide. O'Reilly Media (2009)

A Study of Risk Management in Hybrid Cloud Configuration

Shigeaki Tanimoto, Chihiro Murai, Yosiaki Seki, Motoi Iwashita, Shinsuke Matsui, Hiroyuki Sato, and Atsushi Kanai

Abstract With recent progress in Internet services and high-speed network environments, cloud computing has rapidly developed, with two main forms. First, public clouds are operated by service providers, such as Google and Amazon. Second, private clouds are built by individual companies for their own use. Unfortunately, public clouds have the problem of uncertain security, while private clouds have the problem of high cost. Hence, the hybrid cloud form is now attracting attention, because it offers advantages of both public and private clouds. The problem with a hybrid cloud, however, is management. Using two cloud forms entails a high computational load. Another problem is data handling. For these reasons, risk management in a hybrid cloud configuration is an important issue. Through analysis of risk in a hybrid cloud configuration, 21 risk factors are extracted and evaluated, and countermeasures are proposed and compiled. Although a hybrid cloud involves a wide-ranging set of risks, it offers many advantages. A future subject will be to quantitatively evaluate the effectiveness of the proposed countermeasures.

Keywords: Hybrid Cloud Computing; Risk Management; Risk Breakdown Structure; Risk Matrix.

Shigeaki Tanimoto · Chihiro Murai · Motoi Iwashita · Shinsuke Matsui
Chiba Institute of Technology, Japan
e-mail: {shigeaki.tanimoto,S0942124KM,iwashita.motoi,
    matsui.shinsuke}@it-chiba.ac.jp

Yosiaki Seki
NTT Secure Platform Laboratories, Japan
e-mail: seki.yoshiaki@lab.ntt.co.jp

Hiroyuki Sato
The University of Tokyo, Japan
e-mail: schuko@satolab.itc.u-tokyo.ac.jp

Atsushi Kanai
Hosei University, Japan
e-mail: yoikana@hosei.ac.jp

R. Lee (Ed.): *Computer and Information Science*, SCI 493, pp. 247–257.
DOI: 10.1007/978-3-319-00804-2_18   © Springer International Publishing Switzerland 2013

1   Introduction

Recent years have seen great progress in Internet services and high-speed network environments. As a result, cloud computing has rapidly developed, with two main forms. First, public clouds are operated by service providers, such as Google and Amazon. Second, private clouds are built and operated by individual enterprises for their own use. Generally, a public cloud eliminates unnecessary facilities cost and offers rapid flexibility and scale. For the enterprise user, however, since a public cloud effectively has invisible features in a virtual configuration, the user is uncertain about the cloud's security and aspects of practical use. On the other hand, a private cloud offers visualization of management, since the enterprise operates its own facilities, and guaranteed security according to the company's own policies. The drawbacks to a private cloud, however, include greater cost for maintenance and management of facilities, and so forth [1].

As one example, an incident of missing data and leakage by a cloud operating company, called the "big ripple," occurred in June, 2012 [2]. When a cloud provider's management handles security poorly, serious risks occur only in the public clouds that it manages, so that such incidents may become apparent to users. On the other hand, a hybrid cloud form, combining aspects of both public and private clouds, is now attracting attention. Generally in a hybrid cloud, data requiring high security is handled within a private cloud, while data requiring easy operation at low cost is handled in a public cloud [3].

Thus, although the hybrid cloud form requires maintenance and management of two different cloud forms, its operation also depends on the kind of data. Furthermore, various risk factors, such as accidentally saving to a different cloud during data storage, are involved [4]. For these reasons, it is important to investigate risk management in a hybrid cloud configuration.

In this paper, to understand such risks in a hybrid cloud configuration, we apply a risk management method for analysis and evaluation from a comprehensive viewpoint. The goal is to enable secure, safe hybrid cloud configuration.

2   Hybrid Cloud Configuration

Cloud computing has now shifted to the practical use stage, and many cloud-related services increased sales in 2011. Moreover, many user companies are verifying the possibility and practicality of cloud computing in introducing information and communications technology (ICT). Cloud computing analysis is thus recognized as a key stage in systems configuration [5].

2.1   Reference Model of Cloud Computing

As shown in Fig. 1, software as a service (SaaS), platform as a service (PaaS), and infrastructure or hardware as a service (IaaS or HaaS) are classified as main components of the present cloud computing model. Moreover, in terms of deployment models, cloud computing is classified into public cloud, private cloud, and hybrid or managed cloud. Finally, cloud computing includes the roles of cloud provider and cloud user [6].

2.2 Hybrid Cloud

Although the hybrid cloud appears in the reference model of Fig. 1, its concrete configuration combines a public cloud and private cloud, as shown in Fig. 2. Usually, a company creates a hybrid cloud, and executive responsibility is shared between the company and a public cloud provider. The hybrid cloud uses both public and private cloud services. Thus, when a company requires both public and private cloud services, a hybrid cloud is optimal. In this case, the company can summarize its service targets and service requirements and then use public or private cloud services accordingly. Thus, service correspondence can be attained in constituting a hybrid cloud, not only for a secure, mission-critical process like employee salary processing but also for business such as payment receipt from customers.

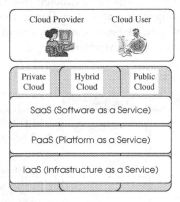

Fig. 1 Reference model of cloud computing [6]

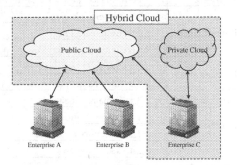

Fig. 2 Hybrid cloud configuration

On the other hand, the main problem with a hybrid cloud is the difficulty of actually creating and managing such a solution. It is necessary to provision the public and private clouds as if they were one cloud, and implementation can become even more complicated. Therefore, since the hybrid cloud concept is a comparatively new architecture in cloud computing and best practices and tools have not yet been defined, companies hesitate to adopt hybrid clouds in many cases [7].

Hence, this paper examines the subject of hybrid cloud configuration in terms of these risks that adoption entails. That is, from the viewpoints of both a cloud user and a cloud provider, we consider what kind of risk is assumed and develop a concrete risk management strategy.

3   Risk Assessment in Hybrid Cloud Configuration

3.1 *Extraction of Risk Factors*

To extract the risk factors in a hybrid cloud configuration, we applied the risk breakdown structure (RBS) method, a typical method of risk management in project management [8]. Table 1 lists the extracted risk factors. As shown in the table, the hybrid cloud configuration was classified at the highest level into system, operation, facility, and miscellaneous categories from a comprehensive viewpoint. A total of 21 risk factors were extracted.

Table 1 Risk Factor List From RBS

High level	Middle level	Low level	Risk factor
1. System	1.1 Software	1.1.1 Application	1.1.1.1 Program distribution
			1.1.1.2 Adoption of duplicate programs
		1.1.2 Data	1.1.2.1 Data distribution
			1.1.2.2 Adoption of duplicate data
	1.2 Hardware	1.2.1 Performance	1.2.1.1 CPU performance
			1.2.1.2 Memory performance
	1.3 Network	1.3.1 Performance	1.3.1.1 Transmission speed
2. Operation	2.1 Public cloud		2.1.1 Resource sharing
			2.1.2 Administration
			2.1.3 Continuity of service
	2.2 Private cloud		2.2.1 Cost
			2.2.2 Talented people
	2.3 Hybrid cloud		2.3.1 ID management
3. Facility	3.1 Public cloud		3.1.1 Administration
			3.1.2 Continuity of service
	3.2 Private cloud		3.2.1 Cost
			3.2.2 Environment
			3.2.3 Convenience
	3.3 Hybrid cloud		3.3.1 Optimum percentage of private vs. public cloud
4. Miscellaneous	4.1 Law		4.1.1 Legal revision
	4.2 Disasters		4.2.1 Disasters

3.2 Risk Analysis in Hybrid Cloud Configuration

Next, we apply risk analysis to address the hybrid cloud configuration risk factors listed in Table 1. Two typical risk analysis methods are the decision tree method and the risk matrix method, which are based on a quantitative and a qualitative viewpoint, respectively [9-10]. In this paper, for the purpose of dealing with security issues, such as privacy protection in a hybrid cloud configuration, we adopt a qualitative viewpoint and apply the risk matrix method. As shown in Fig. 3, this method can classify risks into four categories of risk management (avoidance, mitigation, acceptance, and transference) in accordance with their frequency and severity. These categories correspond with the risk management analysis in the following section. Figure 4 shows an example of a template for deriving countermeasures after classifying risk factors by the risk matrix method.

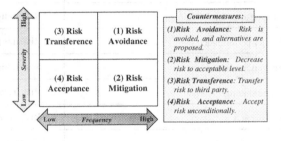

Fig. 3 Risk matrix

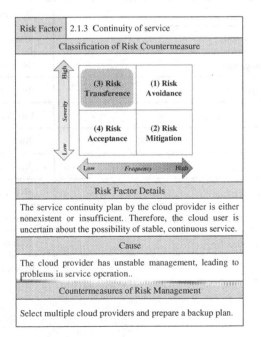

Fig. 4 Example of a risk analysis result

We conducted detailed analysis of all 21 risk factors listed in Table 1, generating the results listed in Table 2. Moreover, Table 3 summarizes the numbers of risk factors placed in each risk management category.

Table 2 Risk factor countermeasures

Risk factor	Countermeasure
1.1.1.1 Program distribution	Risk transference
1.1.1.2 Adoption of duplicate programs	Risk transference
1.1.2.1 Data distribution	Risk mitigation
1.1.2.2 Adoption of duplicate data	Risk transference
1.2.1.1 CPU performance	Risk mitigation
1.2.1.2 Memory performance	Risk mitigation
1.3.1.1 Transmission speed	Risk mitigation
2.1.1 Resource sharing	Risk avoidance
2.1.2 Administration	Risk avoidance
2.1.3 Continuity of service	Risk transference
2.2.1 Cost	Risk mitigation
2.2.2 Talented people	Risk transference
2.3.1 ID management	Risk mitigation
3.1.1 Administration	Risk transference
3.1.2 Continuity of service	Risk transference
3.2.1 Cost	Risk acceptance
3.2.2 Environment	Risk acceptance
3.2.3 Convenience	Risk mitigation
3.3.1 Optimum percentage of private vs. public cloud	Risk mitigation
4.1.1 Legal revision	Risk acceptance
4.2.1 Disasters	Risk transference

Table 3 Number of risk factors in each category

Risk mitigation	Risk transference	Risk acceptance	Risk avoidance	Total
8	8	3	2	21

4 Risk Management in Hybrid Cloud Configuration

In this section we explain each risk factor and its countermeasure in detail according to Table 2.

4.1 Risk Mitigation

Table 4 lists specific countermeasures for the risk factors requiring risk mitigation. These risk factors relate mainly to cloud performance, the ratio of private vs. public cloud implementation, and practical use of the cloud. The countermeasures include the following:

1) Establish accurate estimation methods for data volume and application volume in the cloud.
2) Strengthen employee training for cloud operation.

Table 4 Risk Mitigation Countermeasures

Risk factor	Detailed countermeasure
1.1.2.1 Data distribution	Prepare a data management manual. Upon cloud introduction, educate and train employees.
1.2.1.1 CPU performance	During cloud design, include a significant performance margin to enable efficient cloud usage even when system utilization exceeds estimates.
1.2.1.2 Memory performance	Guarantee sufficient storage capacity to handle cases of excessive system utilization.
1.3.1.1 Transmission speed	During cloud design, properly consider scale, cost, enterprise usage pattern, and so forth.
2.2.1 Cost	Reduce cost by educating employees so that the cloud's operation can be corresponded as much as possible in its company.
2.3.1 ID management	During cloud construction, fully investigate security so as to unify the security control methods of both the private and the public cloud.
3.2.3 Convenience	Private Cloud's operation is made to permeate as an enterprise rule beforehand.
3.3.1 Optimum percentage of private vs. public cloud	Determine a utilization policy for data handling.

4.2   Risk Transference

Table 5 lists countermeasures for the risk factors requiring risk transference. As a general trend, many of these countermeasures deal with the public cloud, such as the following:

1) Join multiple public clouds.
2) Adopt third party surveillance.
3) Use assurances adapted for the cloud.

Table 5 Risk transference countermeasures

Risk factor	Detailed countermeasure
1.1.1.1 Program distribution	Strengthen the management system upon deploying data and programs.
1.1.1.2 Adoption of duplicate programs	Even if the cloud is used mainly on active standby, prepare an additional cloud on cold standby to enable program exchange through manual operation.
1.1.2.2 Adoption of duplicate data	Even if the cloud is used mainly on active standby, prepare an additional cloud on cold standby to enable program exchange through manual operation.
2.1.3 Continuity of service	Select multiple cloud providers and organize backups and other processes in other public clouds.
2.2.2 Talented people	With hybrid Cloud construction, the rule about handling of data is set up. Here, the handling of data is into which Cloud to make the preservation destination of data.
3.1.1 Administration	Deploy multiple public clouds.
3.1.2 Continuity of service	Take out an insurance policy upon public cloud utilization. In addition, request third-party evaluation and surveillance about the cloud provider.
4.2.1 Disasters	Prepare multiple, separate backups for the private and public clouds.

4.3   Risk Acceptance

Table 6 lists countermeasures for the risk factors requiring risk acceptance. These countermeasures deal mainly with legal issues and indirect risks based on external factors. In addition, compromise is necessary for technical problems.

Table 6 Risk acceptance countermeasures

Risk factor	Detailed countermeasure
3.2.1 Cost	Sufficiently investigate the cost of private cloud construction, and ensure that cloud facilities are used efficiently, such as through diversion.
3.2.2 Environment	If a particular situation is judged necessary for the enterprise, approve it in order to develop the business.
4.1.1 Legal revision	Respond flexibly to changes in law.

4.4 Risk Avoidance

Table 7 lists countermeasures for the risk factors requiring risk avoidance. As a general trend, these countermeasures include problems with practical use of the public cloud, such that private cloud usage is optimal. The risk avoidance method of not using a cloud applies only to public cloud operation, but this is not a fundamental solution. The capability of a hybrid cloud to avoid public cloud problems by adopting a private cloud demonstrates the effectiveness of the hybrid approach.

Table 7 Risk avoidance countermeasures

Risk factor	Detailed countermeasure
2.1.1 Resource sharing	Do not use the public cloud but protect the company by using the private cloud.
2.1.2 Administration	If public cloud operation is unsuitable, switch to private cloud operation, and vice versa.

4.5 Summary of Risk Management Analysis Results

Here, we summarize our conclusions concerning countermeasures for risk factors in a hybrid cloud configuration. Table 8 compiles the results listed in Tables 4-7. As seen in the table, there were many risk factors requiring countermeasures for risk mitigation and risk transference. The main tendencies are described below.

Risk mitigation: Many risk factors in this category require advance countermeasures, such as the cloud choice, precise estimation of data volume, and granting of access permissions.

Risk transference: Many countermeasures for risk factors in this category involve the public cloud, as a general trend. Specifically, these include deployment of multiple public clouds, third-party surveillance, and use of assurances adapted for the cloud.

Risk acceptance: Direct problems, such as the scale of an enterprise and its facilities, follow enterprise-level policies. Indirect problems, such as legal revision, require more flexible responses.

Risk avoidance: Practical use of a public cloud entails a data sharing problem with other companies. If such data sharing is unacceptable, the use of a private cloud is optimal, thus demonstrating one benefit of a hybrid cloud configuration.

Table 8 Risk classification results

	Risk mitigation	Risk transference	Risk acceptance	Risk avoidance
System	4	3	0	0
Operation	2	2	0	2
Facility	2	2	2	0
Miscellaneous	0	1	1	0
Total	8	8	3	2

As mentioned above, the choice to deploy multiple clouds and precise knowledge of data volume are important issues for an enterprise introducing a hybrid cloud. Third-party surveillance and assurances adapted for the cloud are also important viewpoints. In particular, synthetic enhancement of not only the technical side but also the practical aspect of the facilities side is important.

5   Conclusion

In this paper, we have analyzed issues of hybrid cloud configuration by applying the RBS and risk matrix methods. In addition, we proposed countermeasures for an enterprise to handle risk factors, enabling operation of a secure, safe hybrid cloud configuration. The 21 risk factors that we extracted indicate that using both public and private clouds in a hybrid configuration requires dealing with a wide-ranging set of problems, both operational and technical. The countermeasures proposed in this paper should help support fundamental risk management for hybrid cloud configuration.

As a future subject, we will consider how to more objectively define the proposed countermeasures. This would allow us to develop a more realistic management proposal.

Acknowledgments. This work was supported by JSPS KAKENHI Grant Number 24300029.

References

1. Nakahara, S., et al.: Cloud traceability (CBoC TRX). NTT Technical Journal, 31–35 (October 2011) (in Japanese)
2. Inoue, O.: First server Failure, Nihon Keizai Shimbun (June 26, 2012) (in Japanese)

3. Microsoft: Shift to Hybrid Cloud (in Japanese), `http://www.microsoft.com/ja-jp/opinionleaders/economy_ict/100701_2.aspx`
4. Tanimoto, S., et al.: Risk Management on the Security Problem in Cloud Computing. In: IEEE/ACIS CNSI 2011, Korea (2011)
5. Goto, A., et al.: The concept for the cloud computing technology CboC. NTT Technical Journal, 64–69 (September 2009) (in Japanese)
6. Uramoto, N.: Security and Compliance Issues in Cloud Computing. IPSJ Magazine 50(11), 1099–1105 (2009)
7. Amrhei, D. (in Japanese), `http://www.ibm.com/developerworks/jp/websphere/techjournal/0904_amrhein/0904_amrhein.html`
8. Risk Breakdown Structure, `http://www.justgetpmp.com/2011/12/risk-breakdown-structure-rbs.html`
9. `http://www.ipa.go.jp/security/manager/protect/pdca/rik.html`
10. NIST Special Publication 800-30 、 Risk Management Guide for Information Technology Systems, `http://csrc.nist.gov/publications/nistpubs/800-30/sp800-30.pdf`

Agent-Based Social Simulation to Investigate the Occurrence of Pareto Principal in Human Society

Khan Md Mahfuzus Salam, Takadama Keiki, and Nishio Tetsuro

Abstract. Agent-based simulation is getting more attention in recent days to investigate the social phenomena. This research focused on Pareto principal, which is widely known in the field of Economics. Our motivation is to investigate the reason of why the Pareto principal exists in the human society. We proposed a model for human-agent and conduct simulation. Our simulation result converges to Pareto principal that justifies the effectiveness of our proposed human agent model. Based on the simulation result we found that due to some factors in human character Pareto principal occurred in human society.

Keywords: social simulation; multi-agent, human-agent model; agent-based simulation, pareto principal.

1 Introduction

As in recent days agent-based social simulation [1] gets mature, new approaches are needed for bringing it closer to the real world. With this purpose, an agent-based model (ABM) was developed for the analysis of how the social structure and group behaviors arise from the interaction of individuals [2]. This model tries to cope with several basic issues by addressing the scenario of agent society but the individual human characteristics are not considered in their model. From that inspiration of their model we try to focus on human individual characteristics. However, as we know human is very complex entity to model initially we consider some specific human characteristics.

Khan Md Mahfuzus Salam · Takadama Keiki · Nishio Tetsuro
Graduate School of Informatics and Engineering,
The University of Electro-Communications, Tokyo, Japan
e-mail: kmahfuz@gmail.com, keiki@inf.uec.ac.jp,
        nishino@uec.ac.jp

R. Lee (Ed.): *Computer and Information Science*, SCI 493, pp. 259–265.
DOI: 10.1007/978-3-319-00804-2_19   © Springer International Publishing Switzerland 2013

2   Backgrounds

2.1   Agent and Agent-Based Modeling

First, in artificial intelligence, an agent is an autonomous entity that observes and acts upon an environment and directs its activity towards achieving goals. Agents can be simple or complex in nature. Complex or intelligent agents may also learn or use knowledge to achieve their goals.

Agents are the "people" of the artificial societies. Each agent has internal state and behavioral rules. Some states are fixed and some states changes as agents move around and interact. All movements, interactions, changes of state depend on rules of behavior for the agents and environment.

In agent-based modeling (ABM), a system is modeled as a collection of autonomous decision-making entities called agents. Each agent individually assesses its situation and makes decisions on the basis of a set of rules as defined for particular environment. Agents may execute various behaviors appropriate for the system they represent — for example, gathering, producing, consuming, or selling. Repetitive interactions between agents are a feature of agent-based modeling, which relies on the power of computers to explore dynamics out of the reach of pure mathematical methods [1]. At the simplest level, an agent-based model consists of a system of agents and the relationships between them. Even a simple agent-based model can exhibit complex behavior patterns and provide valuable information about the dynamics of the real-world system that it emulates. In addition, agents may be capable of evolving, allowing unanticipated behaviors to emerge. Sophisticated ABM sometimes incorporates neural networks, evolutionary algorithms, or other learning techniques to allow realistic learning and adaptation.

2.2   Pareto Principal

The Pareto principle states that, for many events, roughly 80% of the effects come from 20% of the causes. This rule is probably one of the most powerful ideas in the field of social science and economics, which is universally applicable in practically every sphere of our lives. It is also known as the 80–20 rule and the law of the vital few. Even though, *Pareto* originally applied the concept to distribution of wealth (80 percent of Italy's wealth belonged to only 20 percent of the population) and *Juran* applied it to distribution of quality (20 percent of the defects cause 80 percent of the problems), over the years the 80–20 rule has been expressed in a number of different ways. Some of these are as follows.

- 80 percent of the results are achieved by 20 percent of the group.
- 20 percent of your effort will generate 80 percent of your results.
- In any process, few elements (20 percent) are vital and many elements (80 percent) are trivial.

- If you have to do ten things, two of those are usually worth as much as the other eight put together.
- 20 percent of the tasks account for 80 percent of the value.

The numbers don't have to be "20%" and "80%" exactly. The key point is that most things in life (effort, reward, and output) are not distributed evenly. Even if we look at the world economy we find the same phenomena. The distribution of world GDP for the year 1989 is shown as follows.

Table 1 Distribution of world GDP, 1989 [4]

Quintile of Population	Income
Richest 20%	82.70%
Second 20%	11.75%
Third 20%	2.30%
Fourth 20%	1.85%
Poorest 20%	1.40%

3   Building the Simulation Model

Recent days, multi-agent based social simulation models are designed for finding the rules and reasons of the society, to predict the spread out of disease and to find necessary action, or predicting the business process and stock market. The simulation models and its complexity differ based on purpose.

Our simulation model is designed for finding the reasons of Pareto rule in the society. For this we first design a human-agent model. Then we design the simulation environment and conduct simulation to obtain results.

3.1   Human-Agent Model

First, we make a hypothesis that there must be some differences in human characteristics that makes the difference in humans and is the reason behind pareto rule. As according to pareto principal top 20 percent ruled over the society and we assume them as the leader in the society. We consider people like CEOs in companies are successful in terms of earning. A study conducted by IBM showed that, According to CEOs, 'Creativity' and 'integrity' are the two most important qualities for success in business [5]. The study is the largest known sample of one-on-one CEO interviews, with over 1,500 corporate heads and public sector leaders across 60 nations and 33 industries polled on what drives them in managing their companies in today's world.

Based on our hypothesis that is also supported by the findings of IBM study, we assume some qualities or factors made the difference. Initially to make our model simple we assume only one factor (e.g. creativity) is responsible for the making the difference. For that reason, in the simplest form of our model human-agents are created with different creativity score between zero to hundred at random. Zero is the lowest and hundred is the highest motivation factor. We design our human agent in such a way that it interacts with the world based on the creativity score. Also based on their creativity score the have the ability to produce new resources. For each agent resource collection rule is shown by the equation as follows,

$$R_x = \sum_{i=1}^{n} r_i$$

where, x = creativity score, n = number of itaration, r_i = earned resource at turn i

3.2   Simulation Environment

We conduct the simulation by creating the artificial world where we applied following conditions on simulation environment:

- World area is defined as a two-dimensional grid (10 X 10), which consists of 100 location cells. Every cell L (x, y) has some resource point and productivity point.

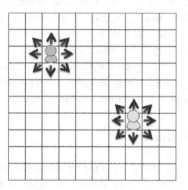

Fig. 1 Describes the environment for the simulation

- Total resource is unlimited for the world. Though initially every location is assigned with some random resources but later human-agents can produce new resources.
- Fifty human-agents are deployed. They scattered randomly and during simulation thy can decide to stay on the same location or can move to eight directions inside the boundary. Multiple human-agents can occupy a single cell at the same time. They can produce new resource based on location agent's productivity factor and self-creativity factor.

Human-agent's rule of behavior is to collect resources from the location they are located in. Their search is local; no agent has any global information and the agents between themselves do not communicate.

3.3 Object-Oriented Implementation

We develop our own simulation system by using object-oriented programming (OOP) language (i.e. Java). OOP languages are natural ones for agent-based modeling. Objects are structures that hold both data and procedures. Both agent and environmental sites are naturally implemented as objects. The agent's represent its internal states and methods are agent's rule of behavior. This encapsulation of internal states and rules is a defining characteristic of OOP and greatly facilitates the construction of agent-based models.

In our system, we have three different kinds of objects and we named those as 'world', 'location' and 'human-agent'. World object collects all the statistics of other objects internal states.

3.4 Simulation

We initialize the simulation as, every cell L (x, y) is assigned with some resource point and productivity point randomly. We also assign creativity factor of every agent at random. Based on the above condition, to study the distribution of wealth in our artificial society we do the simulation in two scenarios—resource gain at random and resource gained based on rule, as follows.

Scenario 1: We run the simulation for random gain. For this run resources are allocated randomly to the agents. For a particular cell, while the resource point comes under ten points new resources are added to the cell randomly.

Scenario 2: Earning resource based on creativity factor resources are allocated based on the creativity factor, i.e., agents having high creativity point can earn more resource than the agents with low creativity factor. We also apply a rule for producing new resource. We set the threshold level as fifty for both creativity score and productivity score. For a particular cell, while the resource point comes under ten point and if its productivity score is over fifty then, new resource can be produced by the current occupied agent if it's creativity score is over fifty. The amount of new resource is a random number. After producing some new resource by a particular human-agent it cannot earn all the resources. We put a maximum limit to half of the generated resources by itself. We perform the both simulation for one hundred iterations and one thousand iterations.

4   Results

In our artificial world agents are accumulating resource measured in units and thus at any time there is a distribution of wealth in society. By looking at every individual agent, we do the calculation to find the distribution of resource in our artificial society.

Table 2 Resource earned at random

Rank	100 Iterations		1000 Iterations	
	Resource unit	In percent	Resource unit	In percent
Top 20 %	840	58%	2205	63%
Second 20 %	392	27%	840	24%
Rest 60 %	214	15%	455	13%

First, Table 2 shows the simulation result for the resource earned at random without considering any feature of the agents. The result shows that, top twenty percent agents owned fifty-eight percent resource after one hundred iterations and sixty-three percent resource after one thousand iterations. Other eighty percent agents own forty-two percent and thirty-seven percent of the resource respectively.

Table 3 Resource earned based on applied factor

Rank	100 Iterations		1000 Iterations	
	Resource unit	In percent	Resource unit	In percent
Top 20 %	65502	82%	43683	89%
Second 20 %	13677	17%	4759	10%
Rest 60 %	717	1%	583	1%

Table 3 shows the simulation result for the resource gain based on creativity factor of human-agents. In this case location property is also applied. We investigate the total resource earned after one hundred iterations and one thousand iterations. It shows that, top twenty percent agents owned eighty-two percent eighty-nine percent and other eighty percent agents own eighteen percent and eleven percent of the resource respectively.

This result indicates that during the random gain though the resource is not distributed evenly it is closer among agents. But when we apply creativity factor our simulation converges to 80-20 principal. Obtained results supports our initial hypothesis that some differences in human characteristics that makes the difference in humans and that is the reason behind pareto principal. The convergence to the pareto principal also indicates our human-agent model's accuracy. Though we should consider other aspects to make our model more realistic.

Furthermore, we closely analyze the top twenty percent agents features and earnings we found almost all the agents have high creativity point. Table 4 shows the detail data. Here we find one interesting phenomena that there is an agent (i.e. agent 3) with low creativity point but gains lot of resource. Which also indicates that in some case things are not fair in the society.

Table 4 Resource earned by top 20 % agents

Rank	Agent ID	Creativity Score	Earned Resource
1	agent 9	89	13371
2	agent 35	89	8736
3	agent 3	12	8580
4	agent 38	63	7403
5	agent 40	84	6754
6	agent 32	73	5782
7	agent 41	64	4433
8	agent 44	83	3799
9	agent 34	74	3663
10	agent 48	85	2976

5   Conclusions

In this research, we investigate the reason of why the Pareto principal exists in the human society. For the investigation we apply multi-agent based simulation technique and we proposed a model for human-agent. We set the simulation environment based on some findings and apply some hypothesis. Our simulation result converges to Pareto principal that justifies the effectiveness of our proposed human agent model. Based on the simulation result we found that due to some factor, i.e. creativity in human character is the reason to occur Pareto principal in human society. Besides our findings, we have two interesting observations.

Observation 1: Sixty-percent of the agents has good creativity score but all of them couldn't earn lot of resources.

Observation 2: In our simulation we found there exist some agent with low creativity factor but it positioned in top 20 on the basis of earning.

Above two interesting observation indicates to some issues— "is there anything obvious that is called luck? ", "does the environment really has effect on human? " — which was not scientifically answered before. By closely looking on the details activity of each human-agent we may find those answers.

References

1. Axelrod, R.: Advancing the art of simulation in the social sciences. Complex 3(2), 16–22 (1997)
2. Epstein, J.M., Axtell, R.L.: Growing Artificial Societies: Social Science from the Bottom Up, 1st edn. MIT Press (1996)
3. Bonabeau, E.: Agent-based modeling: Methods and techniques for simulating human systems. PANS 99(3), 7280–7287 (2002)
4. Human Development Report 1992. UNDP. Oxford university Press (1992)
5. http://www-935.ibm.com/services/us/ceo/ceostudy2010/index.html (last accessed: January 30, 2012)

Modeling Value Evaluation of Semantics Aided Secondary Language Acquisition as Model Driven Knowledge Management

Yucong Duan, Christophe Cruz, Abdelrahman Osman Elfaki, Yang Bai, and Wencai Du*

Abstract. Many theories and solutions have been proposed for the improvement of the learning efficiency of secondary language(L2) learning. However neither an unified view on the functionality of semantics in aiding learning nor an objective measure of the efficiency improvement at theoretical level has been presented in existing literature. This situation hinders the efficient adoption of the semantics aided learning and the explicit planning of a semantics aided learning process. We aim to fill these gaps by adopting an evolutionary strategy towards approaching a holistic solution. Firstly we model the general learning process from cognitive linguistic perspective at the memory level. Then we justify the functionality of semantics aided approach according to specific conditions. Thereafter we propose the quantity measure for the improvement of learning efficiency in terms of reuse level for semantics aided secondary language learning from the perspective of value based analysis.

1 Introduction

The idea of service computing draws a scenario in which services across various types of barriers such as location, time, culture and languages might be connected

Yucong Duan · Yang Bai · Wencai Du
College of Information Science and Technology, Hainan University, China
e-mail: duanyucong@hotmail.com, yang19828@yahoo.com,
    wencai@hainu.edu.cn

Christophe Cruz
Le2i, CNRS, Dijon, France
e-mail: christophe.cruz@u-bourgogne.fr

Abdelrahman Osman Elfaki
Faculty of Information Sciences and Engineering, MSU, Malaysia
e-mail: abdelrahmanelfaki@gmail.com

* Corresponding author.

R. Lee (Ed.): *Computer and Information Science*, SCI 493, pp. 267–278.
DOI: 10.1007/978-3-319-00804-2_20   © Springer International Publishing Switzerland 2013

to serve users' requirements and fulfill business objectives while guided by economical analysis[1]. However there are many challenges which must be conquered
before the idea is fully implemented. For services across multiple language[2] platforms to be composed, it requires an agreed understanding and interpretation of the
service contracts[3] written in different natural languages. This is a big topic in both
natural language processing and involves huge knowledge in linguistics[4]. Many
theories and solutions have been proposed. However since most of these theories
and solutions originate in various research perspectives and target at various solution scopes, very few unification on the opinions of the mechanism of the linguistic
functionality has been achieved among them.

We target at understanding the functionality of the semantics aided approach[5, 6]
in secondary language(L2) learning[7, 8] and quantify the measurement of the efficiency improvement. We adopt an evolution perspective towards approaching the
targets. Firstly we model the cognitive model of L2 learning in terms of memorization. Secondarily we analyze the functionality and application scopes of the
semantics aided approach for L2 learning based on the cognitive model. Finally
enlightened by value based analysis[9], we propose contribution calculation formulas for measuring the efficiency improvement [10] after introducing the semantics
aided L2 learning in accordance with the degree of reuse. The idea of our proposed
evaluation approach might be applied for measurement of a broad scope of knowledge reuse which is not limited to reuse first language (L1) related knowledge for
L2 learning specifically.

The rest of the paper is organized as follows: Section 2 presents the solution
strategy which crosscuts resolving complexity through model driven approach, introducing value based analysis for measurement from knowledge management perspective, and maintain consistency of semantics through evolving conceptualization.
Section 3 presents our research hypotheses which define our targeted problems and
restrict our solution framework. Section 4 lists the fundamental research problems
and constructs a general cognitive process of learning in terms of memorization.
Section 5 formalizes the information foundation of L2 learning and proposes the
value quantification formulas for efficiency improvement of L2 learning. Section 6
presents general implementation steps in accordance to the proposed approach. Section 7 presents related work. Section 8 concludes the paper with future directions.

2 Discussion on Solution Strategy

In this section, we propose the solution strategy of our approach from the dimensions of relieving complexity, quantity analysis and consistency maintenance of
expressions.

2.1 Model Driven Based Strategy on Resolving Complexity

Understanding and improve L2 learning[7] processes is a traditional target for cognitive linguistics. Various theories have been proposed and positive results have been

observed in the past. However most of the confirmed knowledge is limited to certain circumstances under certain hypotheses, so they are partial solutions from a holistic view [11]. We identify that there is an increasingly challenge to manage the complexity of composing related knowledge from a global view in order to harvest their positive contributions while not obligate their application scopes. To composing related L1 related knowledge for the purpose of L2 learning consistently, a premise is that all existing L1 knowledge rules which are to be composed have a complete, clear and uniform description of their application restrictions. However this premise is hardly exist in reality especially in the context of that most of the L1 related knowledge are expressed with informal natural language. From our past experience of resolving complexity through an evolving modeling process in the area of model driven engineering(MDE)[12], we propose to resolve this challenge of complexity through a process of evolving modeling. From an evolutionary perspective of formal semantics [13, 14], we propose an analytical framework to support attaining a global level solution based on the model driven process.

2.2 Value Based Analysis on Knowledge Management Model

Value based analysis has been adopted as a fundamental approach in linguistics [4] which is used to justify the motivation of the evolution of languages from a historical perspective [4]. However the concept of value in [4] is too abstract and too general for guiding specific and concrete analysis at service value level [9] which bridges business analysis and project planning. Firstly we plan to model the intended aiding relationships from L1 learning to L2 learning as an knowledge introduction process from a knowledge management[15, 16] perspective. At the knowledge management level, we can detail the situations where semantic knowledge contribute positively or negatively. Secondly we abstract the application of the knowledge gained from L1 learning to L2 learning as a reuse process. Then we can extend a formal measure of the efficiency improvement in terms of the degree of the knowledge reuse where both the cost and gains of the knowledge application from L1 to L2 can be measured according to economical formulas.

2.3 Evolutionary Modeling from Conceptualization

- *Phenomenon outside*: In general, because of the rich phenomena of language learning targets and material, most of existing solutions will be partial solution if they focus on a specific scenario. But every solution can be confirmed as a contribution as long as the theory is consistent and explains the examples. In another word, individual solutions are bundled with specific application scopes: limited age scope, limited culture scope, certain language, language features of noun, verb, adj/adv, grammar, sentence, etc.
- *Phenomenon inside*: In natural language expression, there are the phenomena of that multiple semantics might be M bundled with a single expression or concept [13].

Phenomenon outside reveals challenges of identifying and expressing the application scopes for the purpose of integrating them to construct a whole solution. *Phenomenon inside* proposes the challenge inside individual expressions if they are not formal. To relieve the disturbance of the understanding of our explanation from these phenomena, we adopt a conceptualization approach [17] to construct an evolutionary explanation framework. In an evolutionary modeling process, the consistency of the semantics of the expression is explicitly maintained while the understanding barriers are supposed to be relieved in the topdown modeling manner where the process of the deepening of the understanding is mitigated from metamodel level to model level.

3 Hypotheses on Secondary Language Learning

L2 learning can benefit from the experience gained through L1 learning for attaining efficiency of both learning behavior/activitiy planning and learning material arrangement.For learning behavior, L2 learning might benefit from the follows:

   (i) The best cognitive experience from L1 learning;
   (ii) The best cognitive experience for general language learning.

For learning material arrangement at semantic level, L2 learning can take full advantage of reusing the semantics structure which has been constructed among language semantics of L1 elements.

We believe that from a holistic view learning material arrangement either dominates learning activity/task [18] planning or is independent from it. So in this work, we focus on laying the foundation for learning material arrangement for semantics aided L2 learning. We propose the following hypotheses as the starting point of further discussion for this purpose:

- *Hypothesis on basic language elements*: the basic notation of a language element and the basic semantics of the same language element are relatively independent existence before the finish of a language learning activity. An obvious fact is that we know basic semantics even before we learn the notations to express them. These semantics have their relatively independent existence.
- *Hypothesis on the form of semantics*: based on the hypothesis of the independent existence of language semantics, we make a further hypothesis that during the first language learning process language elements which are learned by us form several semantic net or semantic structures (SSN) [19]. In this paper, at general knowledge management level, we do not distinguishing syntactic structure and semantic structure in the process of modeling general semantics as to be managed knowledge pieces.
- *Hypothesis on the applicability of reuse*: at learning material preparation level, there are overlaps among the SSN(L1) of attained L1 material and the SSN(L2) to be formed in targeted L2 learning material.
- *Hypothesis on the efficiency of reuse*: Limited to a simplified learning activity cost analysis, reusing SSN(L1) will be cost efficient for learning L2 since that it saves the behavioral effort [7] of constructing SSN(L2).

4 Modeling the General Learning Process

We present the fundamental questions which any related research can not avoid and model the general learning process as a basis for further work.

4.1 Fundamental Questions

Not like empirically employing semantics gained from L1 learning directly as means of improving efficiency of L2 learning, we need to understand the following questions before designing a guided application of semantics aided L2 learning:

- What is the function of semantics in language learning?
- How could its functionality be justified as positive and with what restrictions?
- How could an efficient L2 learning be planed based on knowledge gained from L1 learning?

To answer these questions and ensure the understanding of the explanation can be rendered uniformly to the audiences with clear semantics bundled to intended concepts, we adopt a conceptualization based evolutionary modeling approach in the following sections.

4.2 Modeling the General Learning from Cognitive Perspective

To make things simple, a semantic aided L2 learning process can be modeled as confirming inheritably to the general rules of a general semantic aided learning. In a general semantic aided learning, semantic of a language symbol is any intended meaning which is bundled with that symbol.

Hypothesis on learning through memorization: general language learning is a process of memorization of either static language material or dynamic language production mechanism which turns previously unknown or outside language material or grammar into known or usable language material or grammar inside the mind of a learner. We model the progress of the learning process as with consecutive progressing stages from the perspective of memorization as follows.

1. *Memorization of language symbols* - During this stage, we learn the symbols which is used for representing semantics at existence level for our communication sensor organs such as vision by eyes or speaking/hearing by month/ear, etc.
2. *Memorization of the integration of language symbol + basic semantics of the symbol* - During this stage, we relate our inner semantics knowledge with language symbols to form common concepts which are shared by people in the society.

   feasibility of consistency vs. representation (symbol composition): if it is re quired that a single semantic must be assigned with a unique composition of language symbols, we can imagine that simply from the perspective of learning burden of huge language symbol composition it is not feasible.

3. *Memorization of language symbol + extensive semantics of the symbol* - During this stage, we learn to relate more semantics to basic concepts.

   Multiple semantics formation (concept composition): We model this as an evolutionary process of relating semantics of our expressions composing existing concepts. This stage help us to realize concepts enrichment while avoids the obsession of huge increase of amount of semantic symbols.

4. *Memorization of language elemental concept/word + grammar* - Language elemental concept/word: the integration of language symbol + semantics.

   During this stage, we start to retrospect on the composition of concepts and identified grammar which is the common order of constructing an expression. Grammar helps to unify the expression of semantics while it does not interfere the intended semantic content.

5. *Memorization of language elements emerging from language expressions* - During this stage, we start a journey of language learning based on thought which is on top of existing language mechanism itself. We identify meanings of language expressions as new concepts. This process constitutes an evolution mechanism of a language itself from elementary to higher level. This stage differs from previous stages in the following aspect: observation is changed from "*semantic → concept*" to "*expression → semantic*". During the first 4 stages, our observation relies on the mode of bundling new semantics with language labels to form language elements. In this stage, new semantics is observed from language expressions as a whole. The reasoning process is also supported by our existing language knowledge..

5   Modeling and Evaluation on Learning Efficiency Change

5.1   Modeling and Conceptualization on Multiple Semantics

If a language (LN) is composed as a sequence of M language elements(LE) and the grammar (GR) of it:

$LN(x) = set(LE) + GR$

If a language expression(LEP) is composed as a sequence of $M >= 2$ pieces of LEs following an order:

$LEP(x) = LE(1) → LE(2) → ... → LE(M)$

For every LE(y), the amount of possible semantics which are bundled with it is denoted as pse(y). We denote the possible semantics which are bundled with a LEP(x) as:

$pse1, pse2, pse3, ..., pseN$.

Ideally the amount of the possible integrated semantics of LEP(x) is a Descartes multiply expression:

$pse(LEP(x)) = pse(1) * pse(2) * ... * pse(m)$

From a conceptualization perspective, we can observe that there is an implicit transfer that a LEP expression can become a language element in the form of LEs recursively. A notable thing is that after the connection which forms an expression chain, the semantics of LEP(x) relies on the relationship of composing LEs which is different from the beginning stages where semantics relies solely on the set on of composing LEs and the grammar. We denote this transfer of the essence of the semantics as a transfer from entity to relationship[20] :

$ENTITY \rightarrow RELATIONSHIP$.

If we consider the conceptualization process of an elementary LE as an explicit relationship, it can be reached that at semantics level semantic is relationship:

$SEMANTIC = RELATIONSHIP$

A formal clarification on this formula is as follows:

An explicit explanation from the perspective of the existence of multiple semantics for a single concept [13] is there are more than one semantic bundled with concept(SEMANTIC), concept(RELATIONSHIP) and concept(ENTITY). If we denote a notation of "=" as a binary equity relationship among two concepts of concept(A) and concept(B) iff there is "$n >= 1$" times of shared semantics among the bundled semantic sets of two concepts of concept(A) and concept(B). If we denote a notation of "! =" as a binary equity relationship among two concepts of concept(A) and concept(B) iff there is "$n >= 0$" times of not shared semantics among the bundled semantic sets of two concepts of concept(A) and concept(B). Then the following formulas are founded in the complete background of $< entity, relationship >$ following a description grammar called natural expression function(NEF) [21]:

For concept(SEMANTIC):

$concept(SEMANTIC) = concept(RELATIONSHIP)$
$concept(SEMANTIC) = concept(ENTITY)$

(if we consider concept(CONCEPT) is a form of entity)

For concept(RELATIONSHIP):

$concept(SEMANTIC) = concept(RELATIONSHIP)$
$concept(RELATIONSHIP)! = concept(ENTITY)$

(if we refer the distinction of relationship vs. entity as a rule from a higher model level)

From the complete comparison background of $< entity, relationship >$, we can conclude from above at semantic level:

$concept(SEMANTIC) = concept(RELATIONSHIP)$ for semantics understanding; and $concept(SEMANTIC)! = concept(RELATIONSHIP)$ for conceptual level distinguishing of the two concepts as a whole.

In this paper, when mentioning semantics of semantics aided language learning we refer by default the semantics of a LEP which contains more than 1 LE.

5.2 Justify the Activity of Semantics Based Approaches

We try to answer how in general a learning efficiency improvement can be guaranteed as positive at semantic level?

Hypothesis on linked LEs: the learning effort for a set of individual language elements is higher than the learning effort for the same set language elements which are however interlinked with meaningful/pre-attained knowledge structures. It is formalized as follows:

$LE_{\{individualLE\}} < LE_{linkedLEs}$

In general, semantics aided language learning (SAL) is based on reusing existing knowledge and experience to save the effort of learning language elements.

Knowledge reuse: reuse the static structure in knowledge store which matches the structure of linked language elements.

Experience reuse: reuse the experience of knowledge learning from L1 learning, such as similar words, similar categories, and similar grammar, etc.

5.3 Value Based Cost Modeling at the Level of Reuse

Here we try to model how an efficient learning process can be implemented at knowledge management level. Some of the semantics of individual symbols can be related according to some existing knowledge or existing experiences. According to our hypotheses at Section 3, the explicitly reuse of the existing knowledge from L1 learning will greatly save the learning effort of L2.

We propose to measure the efficiency improvement of L2 learning in terms of value with a demonstrative formula of cost vs. gain calculation as follows:

$IME = LE - RN * ERE$

IME:improvement of learning efficiency

LE: gross learning effort which are spent by a learner without help

RN: times of reuse

ERE: effort of learning the reused knowledge without help.

Derived conclusion:

As long as ERE is positive and $RN > 0$, IME is positive.

5.4 Analyzing L2 Learning According to Cognitive Model

Here we compare the learning process of L1 vs. L2 referring to the 5 stages of the general cognitive model. For L1 learning, all of the 5 stages of the general model can apply. For L2 learning, the difference might depend on the specific approach which are adopted by a L2 learning. Anyway the efficiency of the L2 learning will match either the situation of a form of L1 learning or L2 learning. We can expect the efficiency of L2 learning be no less than that of a L1 learning process normally based on the formula of:

$IME = LE - RN * ERE.$

Under the direction of SAL, we can plan a L2 learning process corresponding to the 5 stages of the general learning model as follows:

For stage 1, the basic notation memorization process of a L2 will be similar to a L1 if there is no linguistic knowledge in which the L1 and L2 can be related to a language family.

For stage 2 and 3, the semantic observation of L2 is by way of semantics of L1 instead of from the result of direct observation with individual senses of a learner.

For stage 4, there are a lot of knowledge which can be reused according to different types of semantic knowledge of word classes. However sometimes there are side-effects ranging from the mismatch of semantics of seemingly similar concepts to inconsistency of seemingly consistent grammar rules[22]. In general, special attention need to be paid to abnormal situations during the process of enjoying reusable knowledge from L1.

For stage 5, a L2 learning has the advantage of that it can benefit from not only its accumulated language learning capability of L2 but also the learning capability of L1. Before the level of a learner's L2 surpass his/her level of L2, he/she has the potential to benefit from the positive difference between his/her level of L1 and his/her level of L2 by using his/her learning capability of L1 to learn L2. The restrictions of usage can be modeled as:

$$level(L1) - level(L2) > 0$$

This restriction is at general level. A specific condition for guiding implementation is that as long as a semantic which is among the possible semantics of a LE in L1 does not exist in the learned capability of a L2 for L2 learner, he/she has the possibility to benefit from using his/her L1knowledge to understand the target language elements of his/her L2. The restriction at semantic level can be modeled as:

$$pse(LEP(x))_{L1} > pse(LEP(x))_{L2}$$

6 Stages of a General Implementation

Based on our previous analysis, we propose the main stages for a SAL learning process consequentially.

1. *Cognitive model.* It describes the underlying process of how human learn a secondary language: the order of learning; the characters of the learning target, the evolution of learning from simple to complex, etc.
2. *Conceptual model.* A conceptual model is based on a conceptualization process. It will set the concept architecture for standardizing the description of the processing and the organization of the learning material in terms of linguistic discipline.The conceptualization process of L2 learning will be different from L1 learning process.
3. *Psychological model* This model is optional but can be adopted for the improving the efficiency of a comprehensive learning by following human learning psychological rules.
4. *Learning model/architecture.* This architecture uses the conceptual model to describe learning strategies which confirm to the facts contained in the cognitive model and psychological model.
5. *Computation model.* Algorithms will be designed and implemented towards goals of information processing in the form of knowledge management or efficiency evaluation according to above architecture with concrete conditions.
6. *Experiments and modification.* Find test cases and use the feedback to refine the proposed theories.

In this paper, we focus on construct the theoretical foundation for the implementation of these stages.

7 Related Work

Secondary language acquisition approaches can be classified according to many categories[23]. From the category of the types of learning targets, most of current existing learning strategies target at learning at vocabulary level [8]. Our proposed approach can be applied not only to vocabulary level but also syntactic level and semantic level. From the category of the types of learning knowledge, most of existing approaches are either based on empirical experiences or on cognitive behavioral models [8]. Since these strategies are mostly based on directly observation or experimental data[24], it is very difficult to extend them into more complex theory to guide more complex targets at semantics level. Also since the scenarios where these approaches are derived are difficult to be related, it is a big challenge to compose related approaches. Our proposed approach supports L1 learning knowledge identification based on reusable pattern identification and knowledge composition centering reuse activity planning. It also supports abstract validation and evaluation in terms of reuse level improvement before experimental test which might be influenced by unknown factors in specific context. Tao et al.[16] emphasis the importance of promoting semantic level learning by ways of knowledge management in E-learning, however the work does neither justify the rationality of introducing semantics nor provides a quantity measure.

- *On cognitive hypothesis*: Cummins [7] has proposed a hypothesis on L2 learning as the common underlying proficiency(CUP). CUP presents that the conceptual knowledge which is developed in L1 learning will contribute to L2 learning as that it can be reused without the need to develop it again. Our work develops a systemic analysis on this view and details related states in a cognitive learning model..
- *On evaluation model* : Saussure [4] proposed that there is underlying value difference which drives the evolution of languages. We focus on the value evaluation representing the efficiency of L2 learning and details the prototype of the calculation formulas which can be extended for modeling general problems in linguistics..

8 Conclusion and Future Directions

Many challenges in information technology can be attributed to fundamental problems at linguistics level. From an evolutionary perspective, we believe that understanding the L2 mechanism is part of the key start point towards building an ideal solution for across language platform information processing such as automatic processing of service contracts in an international background. In view of the absence of theories covering fundamental issues of semantics aided L2 learning at semantic level, we propose a systemic solution which models the cognitive learning process at knowledge management level, introduces quantity analysis of the efficiency improvement at the economical level from the perspective of reuse. The significance of this work lies mainly in that we try to answer the fundamental questions in L2 learning by transferring these problems into other domains where there are familiar solutions through modeling. Specifically we contribute in the following points:

- *formalizing information foundation of semantics aided L2 learning* : from a conceptualization approach [17], we formalize the information foundation of applying semantic aided L2 learning as overlap between SSN(L1) and SSN(L2).
- *modeling L2 learning for value based analysis*: by way of modeling the L2 learning process as a process of knowledge introduction of knowledge management [15] from L1 learning to L2 learning, we successfully transfer an abstract learning process into a domain where value based analysis can be applied directly.
- *quantification of learning efficiency improvement* : we identify the degree of reuse as a key variables/factors for formulating the calculation/measure of the quantity of learning efficiency improvement. We propose demonstrative formulas for applying in L2 learning.

We plan to proceed towards learning activity planning, and refine the proposed approach with comparative experiment[25]. We would also like to use our evaluate approach to identify the cons and pros of existing practices from many sources including empirical and cognitive psychological sources and incorporate them to construct personalized [26] solution.

Acknowledgment. This paper was supported in part by China National Science Foundation grant 61162010 and by Hainan University Research program grant KYQD1242. * stands for corresponding author.

References

1. Duan, Y.: A Survey on Service Contract. In: SNPD, pp. 805–810 (2012)
2. Comerio, M., Truong, H.-L., De Paoli, F., Dustdar, S.: Evaluating Contract Compatibility for Service Composition in the SeCO$_2$ Framework. In: Baresi, L., Chi, C.-H., Suzuki, J. (eds.) ICSOC-ServiceWave 2009. LNCS, vol. 5900, pp. 221–236. Springer, Heidelberg (2009)
3. Duan, Y.: Service Contracts: Current state and Future Directionsmeasure. In: ICWS, pp. 664–665 (2012)
4. de Saussure, F.: Course in General Linguistics. Open Court, Chicago (1998)
5. Dunlap, S., Perfetti, C.A., Liu, Y., Wu, S.: Learning vocabulary in chinese as a foreign language: Effects of explicit instruction and semantic cue reliability
6. Ellis, N C : At the interface: Dynamic interactions of explicit and implicit language knowledge. Studies in Second Language Acquisition 27, 305–352 (2005)
7. Cummins, J.: Language, Power and Pedagogy: Bilingual Children in the Crossfire. Multilingual Matters, Clevedon (2000)

8. O'Malley, J., Chamot, A.: Learning Strategies in Second Language Acquisition. The Cambridge Applied Linguistics Series. Cambridge University Press (1990)
9. Duan, Y.: Modeling Service Value Transfer beyond Normalization. In: SNPD, pp. 811–816 (2012)
10. Norris, J.M., Ortega, L.: Effectiveness of l2 instruction: A research synthesis and quantitative Meta-Analysis. Language Learning 50(3), 417–528 (2000)
11. Duan, Y., Lee, R.: Knowledge Management for Model Driven Data Cleaning of Very Large Database. In: Lee, R. (ed.) Software Engineering, Artificial Intelligence, Networking. SCI, vol. 443, pp. 143–158. Springer, Heidelberg (2013)
12. Duan, Y., Cheung, S.C., Fu, X., Gu, Y.: A Metamodel Based Model Transformation Approach. In: SERA, pp. 184–191 (2005)
13. Duan, Y.: Semantics Computation: Towards Identifying Answers from Problem Expressions. In: SSNE 2011, pp. 19–24 (2011)
14. Duan, Y.: A Dualism Based Semantics Formalization Mechanism for Model Driven Engineering. IJSSCI 1(4), 90–110 (2009)
15. Duan, Y., Cruz, C., Nicolle, C.: Architectural Reconstruction of 3D Building Objects through Semantic Knowledge Management. In: SNPD, pp. 261–266 (2010)
16. Tao, F., Millard, D.E., Woukeu, A., Davis, H.C.: Managing the semantic aspects of learning using the knowledge life cycle. In: ICALT, pp. 575–579 (2005)
17. Duan, Y., Cruz, C.: Formalizing Semantic of Natural Language through Conceptualization from Existence. IJIMT 2(1), 37–42 (2011)
18. Swain, M., Lapkin, S.: Task-based Second Language Learning: The Uses of the First Language. Arnold (2000)
19. Meng, H.M., Siu, K.C.: Semiautomatic acquisition of semantic structures for understanding domain-specific natural language queries. IEEE Trans. Knowl. Data Eng. 14(1), 172–181 (2002)
20. Chen, P.P.: The entity-relationship model - toward a unified view of data. ACM Trans. Database Syst. 1(1), 9–36 (1976)
21. Duan, Y., Kattepur, A., Zhou, H., Chang, Y., Huang, M., Du, W.: Service value broker patterns: Towards the foundation. In: ICIS (2013)
22. Corder, S.: The significance of learner's errors. International Review of Applied Linguistics in Language Teaching 5(4), 161–170 (1967)
23. Stockwell, G.: A review of technology choice for teaching language skills and areas in the call literature. ReCALL 19(2), 105–120 (2007)
24. Dearman, D., Truong, K.: Evaluating the implicit acquisition of second language vocabulary using a live wallpaper. In: ACM HFCS, pp. 1391–1400 (2012)
25. Allum, P.: Call and the classroom: the case for comparative research. ReCALL 14(1), 146–166 (2002)
26. Ayala, G., Paredes, R.G., Castillo, S.: Computational models for mobile and ubiquitous second language learning. Int. J. Mob. Learn. Organ. 4(2), 192–213 (2010)

Author Index